Connecting with Companies

A GUIDE TO CONSULTING AGREEMENTS
FOR BIOMEDICAL SCIENTISTS

Connecting with Companies

A GUIDE TO CONSULTING AGREEMENTS
FOR BIOMEDICAL SCIENTISTS

Edward Klees
University of Virginia
Investment Management Company

H. Robert Horvitz
Howard Hughes Medical Institute
Massachusetts Institute of Technology, Cambridge

COLD SPRING HARBOR LABORATORY PRESS
Cold Spring Harbor, New York • www.cshlpress.org

Connecting with Companies
A GUIDE TO CONSULTING AGREEMENTS FOR BIOMEDICAL SCIENTISTS

Published by Cold Spring Harbor Laboratory Press
Printed in the United States of America

Publisher and Acquisition Editor	John Inglis
Production Manager	Denise Weiss
Production Editor	Rena Springer
Cover Designer	Mike Albano

Library of Congress Cataloging-in-Publication Data

Klees, Edward, author.
   Connecting with companies a guide to consulting agreements for biomedical scientists/ by
Edward Klees and H. Robert Horvitz.
      pages cm
   Includes bibliographical references and index.
   ISBN 978-1-621821-07-6 (hardcover : alk. paper)
1. Consulting contracts--United States. 2. Medical scientists--Legal status, laws, etc.--
United States. 3. Biomedical engineers--Legal status, laws, etc.--United States. 4. Science
consultants--United States--Handbooks, manuals, etc. I. Horvitz, H. Robert, author. II. Title.

   KF905.C67K56 2014
   344.7304'1--dc23

                                                                                      2013050328

10 9 8 7 6 5 4 3 2 1

All World Wide Web addresses are accurate to the best of our knowledge at the time of printing.

For a complete catalog of all Cold Spring Harbor Laboratory Press publications, visit our website
at www.cshlpress.org.

The contents of this book were previously published in 2012 as *Biomedical Consulting
Agreements: A Guide for Academics*.

E.K. dedicates this book to Susan, Jessica, and Rachel.

H.R.H. dedicates this book to the memory of his father
Oscar Horvitz, to his mother Mary Horvitz, and to
Martha, Alex, Joe, and Chris.

Contents

Acknowledgments

We are indebted to the following individuals for their excellent comments and contributions to this book:

Craig A. Alexander
Mark E. Berg
Craig Blackstone
George Q. Daley
Ashley Dombkowsi
Larry Fritz
Nicholas Galakatos
Robert Kester
Monty Krieger
Robert Millman
Darcy M. Norville
Robert Tobey
Yi Zhang

Thanks also to the Howard Hughes Medical Institute as the source of ideas and layout for certain examples used in this book as well as our reference to items from versions of its model consulting agreement.

Ed Klees would like to thank his wife and children for their encouragement and especially for their patience when work on this book ate into family time. Bob Horvitz thanks his family and his many friends who have helped him learn about the world of biomedical consulting. Both authors thank John Inglis, Rena Springer, Wayne Manos, Robert Redmond, and Jan Argentine of Cold Spring Harbor Laboratory Press and Jim Austin of Science Careers, http://sciencecareers.sciencemag.org/.

Disclaimer

This book is intended to give the reader information that is of general applicability. It is a guide about how to think about consulting contracts between academic scientists and companies in the biomedical field. This book is intended to provide only general, nonspecific legal and tax information that might be relevant as of the date of publication under United States law. **It is not legal or tax advice nor intended as legal or tax advice**. This book is not intended to cover all the issues related to the topics discussed. The specific facts that apply to your situation might make an outcome different from what you expect. Every contract reflects the unique facts and circumstances of the parties, including tax status, the subject matter of the contract, and the laws governing the contract. Consult a lawyer or tax advisor whenever specific legal or tax advice or assistance is needed. **This book does not create any attorney–client relationship and is not a solicitation**.

Our opinions are our own and do not necessarily reflect the views of our institutions and employers.

1 | Introduction

WE HAVE WRITTEN THIS BOOK as a guide for faculty in the biomedical sciences who are considering consulting opportunities with industry.

We have frequently observed academic scientists who, although diligent to the point of compulsion when it comes to their research, are indifferent to the contents of the consulting contracts they sign. Whether you fit this phenotype or not, we hope this book will help you see that what appears in a consulting contract in the fields of medicine and biotechnology might matter very much and is worthy of the same careful review as, say, a will or other legal document. We highlight issues that we have encountered and discuss how they can be addressed. We hope this book will convince you that it is worthwhile for you to read and understand any consulting agreement that you sign and to ensure that it accurately describes the parties' mutual expectations.

A Clash of Cultures?

The negotiation of consulting agreements between academic faculty and companies in the biotechnology and pharmaceutical industries often involves a contrast of cultures. On one side is an academic research scientist who is generally accustomed to openness and trust in interactions with laboratory colleagues but who might be impatient with paperwork and is not usually enamored of lawyers and "legal stuff." On the other side is a company that is motivated to protect discoveries and trade secrets and that relies on lawyers in conducting its affairs. On yet a third side is your employer, a university, hospital, or research institute, the affairs of which are more and more prescribed by policies written by lawyers or other risk-sensitive administrators. When a company asks you to consult, not only the interests but also the points of view can be quite distinct.

The goals might be different, too. You might look at the relationship primarily as an opportunity for intellectual interaction, a chance to make a more direct impact on human health and maybe the possibility of earning some extra money, and as such might give little thought to the specifics of the contract. By contrast, the company probably wants your services to improve their products or research and development (R&D) and regards the relationship as a business proposition. And your institution might want to protect your independence as well as its stake in your future intellectual property (IP).

Disparate points of view can meet when a scientist at an academic institution discusses a consulting project with industry. The company operates in the business world, and the business world operates via contracts and the give-and-take of negotiations. In that world, it is expected that the party drafting a contract will write provisions favorable to its interests and leave it to the other party to raise an objection or offer a counterproposal. So, whereas a company will write the consulting agreement to be favorable to its interests, not out of malice but out of practice, you the academic might not know this and indeed might believe that you are expected immediately to sign on the dotted line—or risk offending the company. Furthermore, you might not have the expertise to know which provisions to question and how to modify them to your best advantage.

One might hope that these differences in expectations and experience will have no important implications: The parties will work together well in a stimulating and synergistic way, you will be paid what you expect, and when the time comes the relationship will conclude amicably. But whatever the context, people put contracts in writing so they are protected if something goes wrong. Sometimes, with consulting agreements, things can go wrong.

What Can Go Wrong?

Consider the following situations, which are offered not because they are likely to happen but rather because of the problems that would ensue if they do:

- Your contract does not accurately describe your consulting fee and says nothing about the stock options you thought you were getting. Meanwhile, your company contact—the one who negotiated the terms with you—just cashed out his options, quit the company, and threw away his cell phone. His successor can find nothing in the files about your options.

- The company publishes a paper with your idea. Your laboratory has been scooped.

- The company patents your idea. Your university thinks it belongs to them and sends the company a cease-and-desist letter. The company shows your dean a copy of an IP assignment with your signature on it, and you admit you signed it during a company visit without giving it any thought. The idea is worth millions, and it looks like neither you nor your university will receive royalties.

- You want to consult for another company but your contract bars you from doing so.

- You are a principal investigator for a clinical trial, and you share exciting unpublished data with a friend who works at a company for which you consult. The company buys stock in the pharmaceutical company that owns the drug being tested. The Justice Department announces an investigation of insider trading concerning the pharmaceutical company stock and says indictments are pending. You are served with a subpoena. The sponsoring company calls its lawyers.

- You are awarded stock options, and you exercise them but do not sell the underlying shares. The market price plummets and you are hit with a tax bill based on the price at the exercise date. You cannot raise enough cash to pay the tax by selling the shares at the current price, and you end up owing far more to the Internal Revenue Service (IRS) than you have ever received from the company.

Also, consider these examples:

- In 2011 the U.S. Supreme Court let stand a lower court's decision finding that Roche Molecular Biosystems, Inc. owned an invention made by a Stanford University scientist during a collaboration between the scientist and Roche. The lower court ruled that the scientist validly granted Roche his interest in future inventions arising from the collaboration even though he had already agreed with Stanford to grant his interest in inventions to the University. The court's decision deprived Stanford of the opportunity to collect millions of dollars of licensing fees from Roche, which in turn deprived the scientist of his share of what Stanford would have received.[1]

- In November 2010, the United States arrested a research physician on insider trading charges. The allegation? That the doctor learned about confidential clinical trial results from one company for which he consulted and then passed along the information to another company for which he also consulted. The second company is allegedly an "expert network" firm that collects information from its network of subject matter consultants and then shares what it collects with its customers, which often include stock trading firms. At the same time the U.S. Securities and Exchange Commission (SEC) charged the physician with civil violations arising from the same allegations.[2] This case is at least the third from 2011 alleging misuse of clinical information by a research scientist.[3]

- In 2009 federal regulators began examining insider trading of shares of healthcare stocks. The concern is that many constituencies—including lobbyists, researchers, consultants, and others—have access to nonpublic information about healthcare companies that might be used to trade illegally.[4]

- As the result of pressure from Congress, in 2005 the National Institutes of Health (NIH) began an investigation into alleged improper payments to NIH scientists at grantee institutions by pharmaceutical/medical/biotechnology companies. The investigation showed that several dozen scientists had failed to disclose consulting relationships with companies in these fields. As a result, the NIH imposed new conflict-of-interest (COI) rules restricting ownership of stock of companies in these businesses and prohibiting its scientists from consulting for for-profit companies in these areas.[5] In December 2006, the federal government brought felony charges against an NIH researcher who allegedly consulted for a drug company on work "directly related" to his research without making required disclosure or obtaining proper approvals. The researcher later pleaded guilty to a misdemeanor charge and agreed to community service and to repay the fees he received.[6] Given NIH's status as a bellwether and its ability to define rules for its grantees, its focus on industry consulting as a COI issue has captured the attention of university administrations, many of which began asking whether they are being adequately informed about their own faculty's consulting relationships. More recently the NIH has turned to financial conflicts concerning grantee institutions and their principal investigators.[7]

• In the summer of 2005 the SEC commenced a criminal investigation of physicians who were conducting clinical trials and allegedly sharing unpublished data and trial results with Wall Street firms that had retained them as consultants.[8] According to *The Seattle Times*, which first reported the practice,[9] the Wall Street firms hired the clinicians to get a jump on the market—if the clinical results were strong, the firms could buy the stock of the sponsoring company before the price went up; if not, they could short the stock and make money that way. According to *The Seattle Times*, the clinicians supplied this information in violation of their contractual obligations to the companies that underwrote the trials. Moreover, according to the newspaper, the SEC intended to investigate whether the disclosures were violations of federal insider-trading laws. Clinicians who were targets of this SEC investigation might have faced disciplinary investigations at their medical centers and if so would have expended time and money preparing to defend themselves against the government or the companies that had underwritten the trials.

Needless to say, the enormous increase in consulting by biologists and clinicians over the past two decades has come at the same time that the stakes have gotten higher, and universities, hospitals, and research institutes are paying increasing attention to both the potential problems and the opportunities to strengthen their finances.

Both authors of this book have experience in reviewing and negotiating biomedical faculty consulting agreements, and we have written this book to help guide faculty concerning the ins and outs. The two major messages we hope to convey are simple. First, you should read and understand a consulting agreement before signing it. To sign without reading invites the admonition *Caveat emptor* ("buyer beware")—or, more precisely, *Caveat consiliator,* or "consultant beware."[10] Second, the agreement should accurately reflect what the parties plan to do. There is nothing as important as each party's properly setting expectations for the other.

Contents of This Book

This book discusses issues specific to consulting contracts. We consider issues that are general in nature and also, for those who want more details, particular clauses often included in faculty consulting agreements. We have appended a sample of a basic consulting agreement (p. 121, Attachment A).

It is important to note that we are not advocating a specific set of consulting terms, including those we present here. The resolution of issues in any given consulting agreement depends upon many factors, including who the parties are, their goals and interests, and their respective bargaining positions. What might be important to one person might be trivial for another. For example, one scientist might care greatly what happens to her potential holdings in the case of her death, but another academic without a family or a favorite charity might care little about what happens to his promised shares after he dies. Conversely, one company might desperately need your expertise, whereas another company might not. One company might have ample funding and plenty of stock options; another may be a start-up with little cash. The review and negotiation of a consulting agreement are dependent on the parties and the setting and can be evaluated only by the parties and their lawyers or other professional advisors (for example, in lieu of a tax lawyer, a good accountant can be crucial to evaluating the tax impact of various equity alternatives). There are no "right" or "wrong" positions in a negotiation, and the optimal positions cannot be predetermined. This book is intended to help with identifying issues and is not meant to prejudge any particular outcome.

We stress that the negotiation of a consulting agreement need not and generally will not be adversarial. Rather, both parties can share the goal of using the negotiation to define explicitly in writing their shared expectations concerning the nature of the consulting arrangement. If a negotiation is handled well, the parties can feel closer and more energized. Consider that many of the terms that are advantageous to a consultant are not detrimental to a company, and vice versa. A lawyer can help you evaluate the terms that are important for you. The negotiation should simply be a means of determining and specifying the best combination of terms in the context of the overall interests of both parties. In addition, neither a consultant nor a company wants to begin a consulting arrangement in the wake of a hostile negotiation.

Consulting agreements are by their nature legal documents, and our discussion addresses legal and tax issues. Nonetheless, as stated in the Disclaimer, we stress that **we are not furnishing legal or tax advice**. We encourage faculty to retain an experienced lawyer for advice, particularly for complicated contracts involving potentially sizeable financial interests. Whenever there are complicated issues or tax questions, expert advice is money well spent. To get a recommendation for a lawyer, check with colleagues who have used a lawyer for negotiating their own consulting agreements or ask the faculty union or a local bar association to recommend someone who is experienced with consulting contracts. Guidance also might be available from the university's general counsel or the head of the technology transfer office—and, indeed, you might be required to speak with them about the relationship anyway. As an absolute minimum, get advice from a colleague who has not only signed but more importantly has negotiated multiple consulting agreements.

2 | Issues to Consider When Negotiating a Consulting Agreement

A S IN MANY ASPECTS OF LIFE, setting expectations is vitally impor- tant before commencing on a consulting project. If there are disap- pointments in a consulting relationship, there can be pain, hurt feelings, and anger. Not only could such a situation be uncomfortable, but a nega- tive history also can interfere with future professional opportunities. The goal of any relationship, especially one founded on a legal docu- ment, is to set the right expectations to minimize the risk of disappoint- ment—yours and theirs. Once the expectations are set, they are easier to control. Your goal, as well as the company's goal, is to set expectations.

Doing so does not require an "us-against-them" mentality, which gener- ally is counterproductive. It is better to view the contract process as a time for all parties to ensure they are comfortable with the goals and the terms.

Given the stakes, think carefully. Besides your academic salary, consulting might be your biggest source of income. If stock or stock options are involved, a consulting arrangement can generate much of your net worth. In addition, you are assuming legal obligations, such as the protection of the company's confidential information and the duty to assign intellectual property (IP). The duty to assign IP is significant, because through carelessness you might convey important legal rights to the company and thereby not only violate your duties to the university but also deprive yourself and your institution of valuable IP. (IP issues are discussed further in this book, including in part D of this chapter and in Chapter 8.)

So as part of getting started, ask questions. Speak to knowledgeable colleagues. Look at other consulting contracts. And consider your nego- tiating position: Are your skills unique? How valuable is your expertise to this company?

As noted, consider retaining an experienced lawyer. Consulting agreements are fairly straightforward legal documents, so the expense should not be substantial, and the costs might be tax deductible. It is also possible (for a consultant with great value to the company) to negotiate the contract to include a clause obliging the company to pay your legal expenses. Many professionals have their tax returns prepared by an accountant every year. Very few people draft their own wills. Is your consulting contract any less serious a matter?

KEY ISSUES

Here are some issues to consider in your negotiations with the company leading to a signed contract.

A. Define What You Want in the Contract

You should begin with a list of the terms you believe you have arranged with the company or otherwise would like to be addressed. A natural tendency in reviewing a contract is to respond to what is there rather than to note what is missing. It is far better to make a list beforehand to ensure that important points are not forgotten. Here again it is useful to ask experienced colleagues or a good lawyer what issues seem important to them to help frame what you want to have addressed in your contract.

> *Example*
>
> What will be the specific areas of your consulting? Do you want to exclude any topics so that you can consult for other companies about those subjects or require additional compensation from this company if it later requests help in those areas? Perhaps there are hot areas in your research laboratory you wish to keep out of the commercial realm for now. (This topic is covered in various places throughout this book and particularly in Chapters 4, 7, 8, 9, and 12.)

> *Example*
>
> How do you expect to be compensated, in what amounts, and when? List all elements of compensation you expect to receive, e.g., annual retainer, hourly fee, shares/stock options. If you will receive stock options, how will they vest, and what is their strike price? (Compensation issues are discussed further in Chapters 5 and 6.)

B. The Contract Is a Legal Agreement—Make Sure It Is Right

1. Read the Contract

Unlike many other legal documents, consulting agreements are relatively short with limited legalese. Nonetheless, it is our impression that some faculty members sign them without reading them. Doing so is unwise and risky. The agreement is legally binding and imposes obligations on both parties. You cannot assume that nothing bad will happen because you failed to read the contract. Courts do not recognize an "ostrich defense," except if the contracting party was impaired from acting in a legal capacity.[1] There would have to be rare and unfortunate circumstances for a university professor to have grounds to assert an impairment. In the absence of such circumstances, it would be very difficult to convince a judge or jury that a professor who had offered his or her expertise on some rarified subject acted appropriately by failing to read the contract upon which those services were to be rendered.

> *Example*
>
> You failed to notice that your consulting agreement with a Japanese company states that you agree to the exclusive jurisdiction of the courts of Japan in the event of a dispute. The company refuses to pay you after you put in 150 hours of work on a complicated problem you solved for them. You sue the company here in the U.S., and the company files a motion to dismiss the complaint on the grounds that the contract expressly requires the parties to litigate all disputes only in Japan. (Should you worry about being sued in Japan if the contract is silent on the topic? See Section 14.B.)

So our first recommendation is simple: Read the contract. Also, even if you have consulted for other companies before, do not assume you know it all and can sign new contracts without reading them. Most contracts, even those addressing the same subject matter, are different. Furthermore, new contracts from familiar companies can change. **Read ANY contract before you sign it.**

2. The Words Mean What They Say

As noted in the Introduction, there is a tacit expectation in business that the party drafting a contract will tilt some provisions in its favor. Read carefully. Also, it is not uncommon for the first draft of a contract to include some incorrect terms. It is incumbent upon you to identify and correct mistakes in the text before signing a contract. Courts are reluctant to discard a contract based on the mistake of one party. And again, to the extent the claim might work at all, a judge or jury might simply dismiss the notion that a university faculty member of presumed high intelligence assumed that the plain words of a contract were wrong or justifiably believed the other party did not really mean what it wrote. **Make sure the written terms are correct.**

Example

The contract requires you to assign to the company all IP in the field of your consulting, which happens also to be the field of your university research. The company hears about your newest discovery and sends you a patent application to sign. You also receive a patent application from the university's technology transfer office, so you call your company contact to indicate that you will not be signing the company assignment. The next day you get a letter from the company's patent lawyers instructing you to "sign or face legal action."

This last example is not merely a theoretical risk. The recent *Stanford v. Roche* case[2] involving a test for H.I.V. should serve as a warning to all academics to confirm that their consulting agreements contain reasonable IP provisions that will not undercut their university's rights.

3. If Something Is Unclear, It Might Be Wrong—or Wrongly Interpreted Later

Apart from finding errors, your reading of the contract might turn up provisions that are vague or ambiguous. Often vagueness or ambiguity comes from fast or sloppy work, but sometimes it is introduced intentionally. This situation might arise when the drafting party makes a general statement to avoid dealing with specifics until after the contract is signed or to create wiggle room around an issue in case there is a later dispute.

Example—Ambiguity

Consider the preceding example. The university might say that the IP provision in your consulting contract is ambiguous, because any company should clearly know that a faculty member has obligations to assign inventions to the university, and this assumption is implicit in the language. The company might argue otherwise. In any event, assume the company drafted the IP provision as broadly as possible with the hope of getting a foot in the door in the event your academic laboratory should make a blockbuster discovery in the same area as the consulting services. Maybe the facts underlying an assertion of rights will be weak, but the threat of a lawsuit could work in the company's favor—for example, by inducing the university to offer the company the chance to license the blockbuster before the university shops it to the company's competitors.

Example—Vagueness

The contract says you will be paid $2500 "per day of consulting." One day you visit the company for six hours. Are you owed the full $2500, or something less because the company has an eight-hour business day for employees? What if you work at the company for 15 hours one day? (See Chapter 5.)

Always ask for an explanation of a vague or ambiguous provision and have the provision rewritten appropriately. If that is not possible, at least get an explanation in writing and save it in your files.

4. Legalisms

A contract might contain legal terms that you do not understand. Whenever that happens, you should ask for an explanation and have the language rewritten in a way you do understand.

In general, it is not wise to rely on the company's explanation of an important legal term or clause. For ethical reasons the company's lawyers might not want to speak to you directly to answer your questions. (The company's lawyers are bound to represent the interests of the company.) Instead, they might funnel answers to your questions through your scientific contact, but he or she also might not fully understand the explanation. It is easy for such exchanges to devolve into a game of "telephone" in which what you are told is muddled or wrong. Simply put, do not sign a contract that you do not understand, and engage a lawyer to get clear expert answers to all of your questions.

Example—Legalese

Many consulting agreements contain an indemnity provision. Indemnities, discussed further in Section 14.A, usually obligate the indemnitor (the party providing the indemnity) to pay costs, including legal fees, incurred by the indemnity (the party receiving the indemnity) in case certain events come to pass. These events usually include lawsuits brought against the indemnitee for acts or oversights of the indemnitor. If you are the indemnitor, then you are in essence agreeing to pay the company's damages and legal fees in a suit brought against the company covered by the terms of the indemnity.

5. If It Is Not in the Contract, You Might Not Get It

Your company contact assures you that you will get stock options, but there is nothing in the contract addressing them. What will happen if your contact leaves the company without leaving any records evidencing your claim? If it is not in the contract, then you are at risk of not getting it. There could be any number of reasons why something has been omitted—your company contact forgot to tell the lawyer; the lawyer used another consulting agreement as a template and forgot to add a new clause; the CEO overruled your request, and your company contact was too embarrassed to tell you before leaving the company; etc. Verbal assurances often are not enforceable. **Get it in writing**.

C. Do Not Assume You Have to Sign a Contract as Is

As noted above, you might want to ask questions about and request changes in the contract you are first given. It is not a breach of etiquette to do so. If you have questions, get them answered to your satisfaction. If the wording is wrong, get it fixed.

A contract might come to you looking perfect, but in general contracts require modifications. Do not be surprised if you find things to be fixed. And do not assume resistance—a company might be perfectly willing to make all of the changes you desire. But you cannot get changes you do not request. Although you might not get all the modifications you want or even obtain clear answers to all of your questions, at least you will know where you stand after an appropriate discussion.

It is important and expected that you focus on issues that matter to you and that you try to identify all of those issues as early in the negotiation as possible. Doing so can prevent unnecessary back-and-forth conversations that might become frustrating (and can be expensive if lawyers are involved) and might lead a company simply to decide your consulting is not worth the effort.

Occasionally a company will say that its contract is nonnegotiable. Sometimes such statements are themselves negotiation tactics and you can get changes made, especially if the company needs you. A lawyer can be helpful in this situation by giving you perspective on what your real bargaining power is and in developing a convincing argument for why your changes are necessary.

Although it is always appropriate to ask for clarifications and explanations, be careful to calibrate your efforts to change key terms, such as your cash and noncash compensation. Consider the extent to which your consulting work will be critical to the company and if your skills are unique. Always be ready to justify any request for more compensation. Do not assume that the company has ample excess cash or plenty of stock options to issue. The more you can understand the company's financial position and your importance to the company, the better you can fashion a request that will be favorably received. On the other hand, if you do not ask for additional compensation, you will not get it.

D. Protect All Intellectual Property

As an academic, you are accustomed to an open exchange of ideas. In dealing with commercial entities, though, it is imperative to protect all IP that does not belong to them. Such IP includes not simply your own ideas and discoveries. Any IP you have obtained from others—whether from a laboratory next to yours, a collaborator at another institution, or other third parties, including other companies—should not be shared with, or committed to, a company that has no rights to that IP. This caution includes not granting inappropriately broad rights to a company under your consulting agreement or assigning to the company a patent or IP rights that are drafted too broadly.

We devote Chapter 8 to a closer analysis of IP rights.

E. Review University Policies

Do not overlook the fact that most universities, hospitals, and research institutions have policies that explicitly address consulting issues. These policies cover conflicts of interest (COIs), and many include specific consulting policies, sometimes called "conflict of commitment" policies. Note also that public universities and hospitals might have stricter policies than private universities or research institutes and that recipients of government (e.g., the National Institutes of Health) funding also must comply with any applicable rules.

Of the policies we have reviewed, many require advance approval of all faculty consulting agreements, whereas others require disclosure only afterwards. It is fairly universal for universities to set at least some general rules about outside consulting, including time limits and restrictions on certain activities. It is thus incumbent upon you to check your institution's policies.

In addition, the laws of many states address COI issues relating to faculty of their public universities and federal laws apply to recipients of U.S. government grants. In some situations, a clinician's consulting activities might be subject to a set of complex conflict of interest laws under federal and state Medicare and Medicaid rules, and perhaps even state medical licensing rules. In these circumstances the clinician's hospital or university may be involved in the approval and documentation of her consulting arrangements. We do not address this topic except to note its potential importance to physicians and others who treat patients. Often you can find information about relevant federal and state laws on the university's policy homepage.

Although we refer here to universities, the same principles apply to hospitals and research institutions.

Many university COI policies cover the following:

1. Time Commitment

Most of the policies we have seen set a maximum amount of time that faculty may devote to consulting. The upper limit might be as generous as one day a week, but some institutions have a lower limit, such as 36 days or even 24 days a year. At least one prominent university requires forfeiture of compensation earned in violation of the time restriction, so you should be sure that the time commitment in the consulting contract does not exceed what the university permits.

2. Disclosure of Laboratory Results

Many policies prohibit faculty from disclosing nonpublic results from their laboratories or other sensitive information to companies and others. This issue is discussed in more detail in Chapters 4, 7, 8, 9, and 12. As already noted, the issues here are not limited to what is permitted under university policies. Disclosure of confidential information relating to commercially valuable results in your laboratory or those you learn from other laboratories also might raise serious legal risks. (See Chapters 7 and 8.)

3. Use of University Resources

Many policies prohibit the use of laboratory or other university facilities for outside consulting. This issue might arise, for example, if the consulting contract calls for meetings at your laboratory.

4. Involvement of Postdoctoral Researchers and Students

Some policies specifically ban the involvement of one's own trainees in consulting projects. Others do not permit consulting in general by students, postdoctoral researchers, and other mentees.

Regardless of policy, assigning trainees to one's private consulting work might not be a good idea. Laboratory members who are excluded from the work could become jealous or insecure, and those who are included might question the relevance of the work to their education or whether the amount they are being paid, if any, is adequate. Laboratory members also might misunderstand your confidentiality obligations.

5. Conduct of Research

Your institution's policies might not permit you to conduct research for third parties. Again, such activities might not be a good idea anyway, especially if the paid research could branch into your academic pursuits. (See Chapters 8 and 9.)

6. "Significant Financial Interest"

Many institutions address consulting compensation. Some prohibit a consultant from having a "significant financial interest" or "significant equity interest" in a consultee company, whereas others can permit such an interest after review and approval in specific circumstances. What constitutes a significant interest varies. Some institutions say that a 5% equity stake in a company constitutes a significant interest. In addition, we are aware of at least one institution that requires special institutional approvals if cash compensation will exceed set thresholds.

7. Employment by Company

There could be tax or other reasons why you might want to be a company employee instead of a consultant, including the opportunity to receive a certain type of stock option that offers tax advantages to corporate employees (see Chapter 6). However, a number of universities do not permit outside employment, and others require institutional approval in limited circumstances. Check first.

While on this topic, we note that there is a California law that might deem a consultant to be a company employee if the company has the contractual right to treat the consultant's IP as work for hire. (See Section 8.C.)

8. Multiple Relationships with Company

Some institutions will not allow you to engage in certain types of activities with a company if you are also a consultant to that company. For example, some do not permit you to accept sponsored support or gift funding from a company while consulting for the company or to serve as an officer or director of a start-up. In any event, multiple relationships with a company can create complex challenges in keeping obligations under one arrangement separated from the obligations under the others. This concept is discussed further in Chapters 7–9.

9. Use of Name or University Letterhead

You might need institutional permission for the company to use your name or the university's name in promotional literature or otherwise, whereas some universities flatly prohibit promotional use of the university name by consultee companies. Also, be mindful in case your university says that faculty letterhead cannot be used for consulting work. Consider this issue if you plan on sending letters on university letterhead for the company's benefit, such as recruiting other academic colleagues for the company's Scientific Advisory Board (SAB).

10. Consulting during a Sabbatical

Some institutions prohibit consulting during a sabbatical on the grounds that the purposes of a sabbatical are inconsistent with active consulting. Sabbaticals can be contrasted with leaves of absence, about which we have seen no such restriction. One explanation for the distinction appears to be that the university continues to pay salary to faculty who are on sabbatical, whereas those on leave typically are not on the university payroll.

3 | What Constitutes Consulting?

A CRUCIAL QUESTION IS WHAT EXACTLY is meant when one speaks of consulting. For your own protection, you should think broadly. Certainly many university policies do so. Stanford University's policy, for example, states:

> In general, consulting is defined as professional activity related to the person's field or discipline, where a fee-for-service or equivalent relationship with a third party exists .... The principle is that, in consulting, a person agrees to use his or her professional capabilities to further the agenda of a third party, in return for an immediate or prospective gain.[1]

Thus, as far as your institution is concerned, you might need to comply with university policies whenever you are to receive fees from a company.

Accordingly, consulting might include, for example, service on a corporate Board of Directors (BOD) or an SAB, founding a start-up, a presentation or attendance at a company seminar, acting as an expert witness, or giving advice to a venture capital firm. For these reasons, the most prudent course is to assume that all company-related services might trigger compliance obligations and check to determine if your situation is covered by your university's policies. If any potential conflict exists, address it directly. If in doubt, call the dean, the university technology transfer director, or the general counsel's office for interpretation or advice. Do not put your university status in jeopardy by failing to consider the university's policies.

At a minimum, consider asking for the insertion of a statement in your consulting agreement that as a university employee you are subject to its policies in place from time to time.

Bear in mind, though, as with all protective language in any contract, adding this provision will not insulate you from problems that might arise later if you do something inconsistent with it. Ultimately it will be your actions that will be determinative. Actions that violate contractual protections can render them moot. If you are asked to do something outside the terms of the contract that might violate university policies—for example, signing an IP assignment that might override your university obligations—say no or ask for intercession from the dean or the technology transfer office, or ask a lawyer for help.

Even if there is no relevant university policy, we think that it is a good rule of thumb to apply the principles of this book to any document a company asks you to sign for which you will be paid or incur IP or confidentiality obligations.

The following relationships generate consulting issues addressed in this book.

A. Service on an Advisory Board

When considering service on a company's advisory board, such as an SAB, the description of the board's areas of responsibility might by necessity not be specific and rather only illustrative ("The areas of SAB responsibility shall include, without limitation...."). Presumably the amount of compensation for SAB service will reflect the greater breadth of responsibility and possible decreased flexibility for you to pursue other opportunities. Nonetheless, you should consider the extent to which the agreement can expressly exclude areas in which you owe or might wish to convey obligations to others. For example, the agreement could list research areas that will not be addressed by the SAB, or specify areas that will be handled by a subcommittee of the SAB from which you will be excluded or that will be discussed only after you excuse yourself from the discussion.

Also, be mindful of the extent to which the SAB will engage in actual research for the company and whether you are comfortable with that aspect of the job and whether it would be acceptable under your institution's policies. See the next section, "Conduct of Research."

B. Conduct of Research

There might be situations in which the company expects you to conduct research as part of the consulting arrangement. There are at least two issues to consider before agreeing to do so. First, as discussed in Section 2.E above, you may run afoul of your institution's policies; check first. Second, assuming the project is in the same field of your university research activities, consider the extent to which the company work could become intertwined with your university research. It is possible that the company's publication restrictions and IP rights under the consulting agreement could impinge upon your right to publish research results or create issues with your university under your duty to assign discoveries to it. See Chapter 12, "Multiple Relationships with One Company."

C. Meetings with Investors

Smaller companies, especially start-ups, sometimes will ask "star" advisors to speak to prospective institutional investors or meet with stockholders. If the company has this intention, you might want to understand and address its expectations concerning these meetings. For example: What role are you to play? What kinds of things are you expected to say? How will your relationship with the company be identified? This discussion can occur informally, but if the contract imposes an explicit obligation on you in this regard, make sure it reflects what you are comfortable doing. In this context, also consider your university's conflict-of-interest and use-of-name policies (see Chapters 7 and 8).

If the company expects you to explain or even promote its business and research, bear in mind that all company spokespeople can be held liable under antifraud or securities laws if information they furnish is later alleged to be misleading or wrong. Although we are not aware of any scientific consultants being sued for what they told investors, there is no reason that consultants should accept a role as a company spokesperson without obtaining protection similar to that which the company offers to its officers, employees, and directors who play a similar role.

There are usually two forms of protection that can protect the consultant. First, depending on whether you are a company insider, you may be covered by the company's indemnity of its BOD and management under its charter. If not, the consulting contract can include a company obligation to indemnify you for these activities. Second, if you are a company insider, you might qualify for coverage under the company's Directors and Officers (D&O) insurance policy. If not, then (although unlikely) the company could purchase a similar policy for you, or (more likely) you can take out such a policy and ask the company to adjust your compensation upward to offset some or all of the cost. Your personal umbrella insurance policy might also provide coverage depending on its terms. Nonetheless, there usually is no indemnity or insurance coverage for intentional violations of law or one's gross negligence.

Note that a company's indemnity of the BOD and management under its charter might be inadequate and/or revocable. Similarly, D&O insurance might not be as robust as one might desire, and a company might not have a legal obligation to renew it (or to continue to include you if your insider status should lapse). Furthermore, in the current economic climate it is not safe to assume that the underwriter providing D&O (or umbrella) insurance is or will remain financially sound. Check all of these issues carefully before you agree to perform duties for a company that will expose you to external constituencies, such as acting as a company spokesperson. This area is one in which a lawyer and an insurance agent could be particularly helpful.

D. Serving as a Company Director or Officer

A company—perhaps a start-up you have helped to found—might want you to serve as a member of the Board of Directors or as an officer (e.g., a chief executive officer (CEO), a president, a chief scientific officer (CSO), or a vice president). There are several issues to consider before saying yes.

First, check your university's policies. As noted, our sampling suggests that some institutions prohibit faculty from acting as a corporate officer, and at some universities serving as a director is restricted. In any event it is common to require faculty to disclose the position in advance or on some regular basis.

Second, a director or officer owes a fiduciary duty to the company. This duty is understood to impose a legal obligation to devote whatever time and effort as are necessary to prudently and faithfully discharge obligations to the company.[2] The indeterminate nature of commitment in terms of time and effort cannot be avoided in advance by contract. Serving as an officer or director of a company with ambitious objectives (such as an initial public offering, or IPO) can require much more time than you might have anticipated. In addition, if your university sets a maximum for consulting time, you might find yourself legally obligated to exceed that limit.

Third, and very important, if something goes wrong, directors and officers can be sued by shareholders, third parties, and even the company itself. A lawyer can examine whether your indemnity protection from the company is sufficient and irrevocable, and an insurance advisor can confirm that the company has adequate D&O insurance in place and that this insurance would cover you.

Next, subsumed within a director's or officer's fiduciary duty is the obligation to offer potential "corporate opportunities" to the company before offering them to others. The corporate opportunities doctrine generally requires a director or officer to offer commercial ventures or ideas to the company first before using them herself or offering them elsewhere.[3] As such, it seems to us that this doctrine could include a duty to offer the company the first right to new inventions made in your laboratory. Consistent with our view that the goal of contract negotiation is to set expectations, you should examine the extent to which your fiduciary duty to the company could come into conflict with your university's IP policies and your own interests.

Finally, if the company is a public company or is about to go public, a director or officer is subject to important federal laws, including those relating to corporate governance and insider trading. Both of these topics are outside the scope of this book. In brief, in the wake of Enron and other scandals, Congress passed the Sarbanes-Oxley Act, which imposes greater obligations of oversight upon officers and directors of a public company with significant criminal and civil penalties for violation.[4] Also, in addition to the laws prohibiting anyone from engaging in insider trading,[5] there are securities laws applying specifically to officers and directors of public companies. Among them is a law requiring an officer or director to publicly report to the SEC all personal transactions in the company's stock or option exercises,[6] as well as another law (sometimes called the "short-swing profit rule") that requires an officer or director to give up all profit on her "matching transactions" (i.e., a purchase and sale, or a sale and purchase) in the company's stock made within a six-month period.[7] The company also is required to report to the SEC and its shareholders if an officer or director fails to timely file its trading reports or engaged in short-swing transactions in the company's stock.[8] The SEC's rules concerning stock transaction reports and short-swing trades are complicated and contain traps for the unwary. If you decide to become a director or officer of a public company, first make sure you are briefed about your obligations under the securities laws.

In light of these issues, some company founders have successfully arranged to be "board observers" rather than members of the company's Board of Directors. Board observers can be contractually granted the right to attend, but not to vote at, Board meetings and to get copies of all Board materials.

E. Advising Venture Capital Funds

Our discussion so far assumes that the consultee company is seeking your advice about its own R&D. This role is a traditional one for academic consulting in the biomedical sciences. However, more and more biotechnology venture capital (VC) funds are retaining academic scientists as consultants to examine the technology platforms of companies in which the VC funds might invest. Consulting for VC firms, which is in some ways similar to and in some ways different from consulting for biotechnology and pharmaceutical companies, is largely outside the scope of this book. Consultant compensation by VC firms can include a portion of the firm's potential gain on its investee companies (called the "carried interest" or the "carry"). So, unlike receiving stock options in an ongoing business, you get a profit interest in the VC firm itself and hence a profit interest in the firm's portfolio of companies. This topic is discussed in Section 6.I. We suggest that if you are considering consulting for a VC firm you seek experienced advice concerning your agreements.

F. "Expert Network" and Hedge Fund Consulting

We noted in the Introduction how a research physician was recently accused of communicating confidential clinical trial information to an "expert network" for which he consulted. Expert networks are firms that hire, among others, academic scientists in the position to share cutting-edge information with the network's clients. Some of these clients are investment firms that might use this information to bet for or against stocks. Therefore, it is possible that an expert network might be financially motivated to tacitly encourage or turn a blind eye to the improper disclosure of confidential information by its scientific and other advisors. For this reason, expert networks have caught the attention of U.S. prosecutors and the SEC. Those who watch the SEC and the courts expect new and possibly broader rules to be adopted soon to address the activities of expert networks and the investment firms they work for. As noted below, Massachusetts has already started to take action in this area.

We suggest that before you agree to consult for an expert network, a hedge fund, or other investment firm, you conduct due diligence about whether the company might be seeking inappropriate access to confidential information and that you adhere scrupulously to all laws and university policies about sharing confidential or nonpublic information.

In thinking about this topic, it would be unwise to assume that you have the skills and omniscience to ascertain what would be valuable information for an investment firm and are capable of disclosing only confidential data that the firm would find insignificant. What might appear to be trivial could prove to be a missing piece of a much grander investment analysis. Indeed, investment firms seek an edge by amassing disparate pieces of information that their competitors cannot obtain or do not understand (so-called mosaic analysis). To start evaluating the merits of confidential data or other information is a slippery slope, and an academic scientist might not know how a piece of information fits into the expansive range of knowledge possessed by investment firms that earn their millions from gathering better information than the rest of the market.

If you are comfortable that the expert network or investment company shares your ethos, it then becomes important for the contract to confirm the understanding. But the true test, as always, is what actually occurs within the consulting relationship after the ink on the contract has dried. Indeed, as alluded to above, one frequent criticism of expert networks is that they do not adequately monitor or constrain the topics of conversation between their experts and the consultee companies despite contractual restrictions to this effect.[9] In response, in late 2011 Massachusetts enacted a law requiring advance disclosure to investment firms of all subjects that an expert is prohibited from discussing with them.[10]

G. Expert Witness Services—Litigation

This topic is also beyond the scope of this book. We note briefly several issues to be considered with respect to expert witness services provided in connection with litigation. Note that "expert network" consulting discussed in the prior section is different from expert witness services. Also note that court rules relating to expert witnesses vary across jurisdictions, and the rules within a jurisdiction can vary depending on whether the expert witness is expected to testify or only to provide non-testimonial assistance.

The first issue is the litigation-related burden. If you are retained to appear in court as an expert witness, the opposing party is entitled before trial to obtain information about your qualifications to testify as an expert. Depending on the relevant rules, the opposition can be permitted to make a "discovery request" requiring you to produce certain kinds of documents relating to the subject of your expertise; "interrogatories," or a set of questions about you and your qualifications that must be answered in writing; and/or a "deposition" in which you will be questioned in person by opposing counsel before trial. It is not unheard of for opposing counsel to make burdensome requests as part of the gamesmanship of litigation. To someone unaccustomed to expert witness work, the attention to these requirements can be onerous if not rancorous or annoying.

The second issue is time commitment. As suggested above, the time involved with expert witness work often encompasses more than simply testifying at trial for a day or two.

Finally, although it is uncommon, your university might limit expert witness work by faculty. As the employers of faculty members serving as expert witnesses, universities might be drawn into the litigation in certain jurisdictions by being served with discovery requests and interrogatories relating to your work or qualifications. For this reason, an institution might want your agreement with the company or its law firm to cover any costs that the institution might incur in connection with the litigation.

It is worth mentioning that sometimes overlooked by academics beforehand is the reputational risk associated with expert witness work, especially when the work might attract public attention in an unfavorable light. This issue also might raise concerns at your institution.

H. Company Seminars, Speeches, Symposia—CDAs

We address general issues about confidentiality in Chapter 7 and in section I of this chapter. This discussion relates specifically to the document you often will be asked to sign before speaking at a company event or attending a company meeting.

If a company invites you to present a lecture or attend a company seminar, meeting or symposium, the company might send you a "confidential disclosure agreement" (CDA)—sometimes called a "nondisclosure agreement" (NDA) or simply a "confidentiality agreement"—to sign before you get there. The purpose of the CDA is to protect the company from your misuse of their proprietary information that you might learn during your visit. We know academic scientists who refuse to sign CDAs on the basis that they do not want to learn anything confidential at the company and the company itself should not want to tell them anything of that nature. Some universities require prior review and approval of CDAs proffered to faculty for this or other purposes.

Too often these CDAs are written very broadly. For example, instead of referring to the specific purpose of the CDA—say, to present a lecture—the agreement will recite that the CDA is in connection with "possible business opportunities" or words to that effect. The CDA also might not state the period of time during which confidential information might be communicated or the "tail period" after the agreement is signed during which you will remain subject to the nondisclosure obligation. Therefore, the company could take the position that you remain legally obligated to keep their information confidential for the rest of your life.

Here is how a typical CDA might describe what you must keep confidential:

Example

"This Agreement covers the disclosure by the Company to Dr. X of certain information of a confidential or proprietary nature ('Confidential Information') in connection with the discussion of a possible business transaction. Dr. X shall maintain the Confidential Information in confidence."

There are several obvious problems with this language. First, it is vague about the purpose of the discussions and the types of information that might be shared. Second, it does not indicate how long the discussions will likely continue. There is no reason not to be more specific, if only to confirm the parties' understanding about the nature and extent of their discussions. Such specificity also helps establish walls between this agreement and any other arrangements you have or might enter into with the same company. This way, there is less likelihood of misunderstandings as to your work or obligations under each arrangement.

Here is how the CDA might be made more specific:

Example

"This Agreement covers the disclosure by the Company to Dr. X of certain information of a confidential or proprietary nature ('Confidential Information') in connection with the seminar to be presented by Dr. X at the Company on or about December 1, 2011 relating to alpha omega signaling pathway biology and alpha omega expression research as well as discussions with Dr. X after the seminar on such date relating to the Company's research program involving the alpha omega pathway. Dr. X shall hold in confidence all Confidential Information that is marked prominently as confidential upon first presentation to Consultant, or if first presented to Consultant other than in writing, is reduced to writing and marked prominently as confidential and presented to Consultant by no later than December 10, 2011. Dr. X's obligations hereunder with respect to holding such Confidential Information in confidence shall expire on December 1, 2013."

The foregoing discussion applies to CDAs relating to corporate visits in connection with such events as a speech or seminar. CDAs can arise in a variety of other contexts, however. For example, a company that is interested in your consulting services might first want you to sign a CDA so the company can show you their technology to allow you to assess how well you might be able to advise them about it. In this circumstance you should consider the same points identified above. You also should consider whether the CDA addresses certain specific items such as (1) appropriate definition of "Confidential Information" (see discussion in Chapter 7); (2) a reasonable tail period; and (3) a reasonably specific summary of the purpose of the CDA. In addition, if you do decide to consult after signing the CDA, make sure that the consulting agreement says that it will replace the CDA and therefore that the CDA is no longer effective. You thus put all your continuing obligations into the consulting contract and need not ever refer again to the CDA. This way, there will be no potential confusion over which agreement (CDA vs. consulting agreement) will govern your ongoing services.

Example

[Insert in CDA:] "This Agreement shall expire upon execution by the parties of a definitive consulting agreement, if any."

Another alternative is to have the consulting agreement include a statement to the effect that the CDA is canceled.

Example

[Insert in consulting agreement:] "This Agreement replaces and supersedes in its entirety the Confidential Disclosure Agreement between the parties dated as of December 1, 2011 (the 'CDA'). The CDA is hereby terminated in all respects."

For more about confidentiality obligations in consulting agreements, see Chapter 7.

As noted, CDAs can arise in many other contexts. For example, either the company or your institution might require a CDA before you discuss a potential collaboration with the company or a possible license of your laboratory's IP that is related to the company's technology platform. Your institution might handle these other kinds of CDAs differently from the manner we discuss here. If you get a CDA for a different subject matter and are unsure what to do, contact your technology transfer office or university counsel.

I. A Note about CDAs Linked to Consulting Agreements

Sometimes a company will choose not to put the consultant's confidentiality obligations in the consulting agreement but instead will furnish a separate CDA. We find this approach to be awkward and unnecessary. First, it sets out the terms of the agreement in two legal documents, and with this process comes the risk that the terms are inconsistent with each other. Second, to be prudent, you or a lawyer would need extra review time to ensure that the two documents are consistent—for example, what if the consulting agreement excludes university research from the definition of "Consultant Inventions" and the definition in the CDA does not?—and otherwise use the same definitions and terminology.

If a company proposes using a CDA, we recommend asking that the terms of the CDA be integrated into the consulting agreement. Often companies will agree to do so.

J. A Final Word Concerning Chapter 3

So far we have suggested some things for you to think about in considering a new consulting relationship. Now, assuming you have discussed all the issues with the company, it is time to look at specific issues that typically appear in consulting agreements. Where possible, we try to delineate a range of positions that can be considered in response to terms that can be more heavily negotiated.

We note two points. First, as stated already, our goal is to educate rather than to lobby for a particular result; as already noted, we are not providing legal advice. Second, the discussion to follow is by nature a summary. Do not assume that one size fits all or that we have covered all issues, or that the issues we have covered are discussed exhaustively.

4 | Scope of Services

EVERY CONSULTING AGREEMENT SHOULD state the nature of the services to be provided. This provision can serve your interests if it is used to exclude sensitive areas of research or preserve areas of expertise for later negotiation with this or another company. And as stated in the Introduction, by setting the company's expectations, you can avoid conflict later.

There are usually two relevant provisions. The first describes the "Field" of consulting. This clause offers an opportunity to tailor the description of the "Field" to the specific field of expertise being offered by the consultant. The more specific and limited the Field, the more flexibility the consultant retains.

> **Example**
>
> Dr. Smith is an expert concerning angiogenesis in mouse tumor models. Company X has submitted a consulting agreement stating "Dr. Smith shall advise Company X in the field of oncology." Better (and more accurate) possible descriptions could include "Dr. Smith shall advise Company X in the field of angiogenesis in mouse tumor models" or "Dr. Smith shall furnish advice concerning the company's research program in the field of angiogenesis in human cancer with reference to mouse tumor models."

There are some experts who would advise not to narrow the field too much, because the consultant would lose opportunities to help the company. We agree that it makes no sense to set a "Field" in an impractically tight manner or to focus on this issue unduly if your consulting duties in the Field are not "exclusive," that is, the contract says you may furnish consulting services to other companies in the same Field. But we believe that more is to be gained by making sure that the Field is defined sensibly. If it turns out that the contract is not written sufficiently broadly, then the parties can always amend it later. But once the consulting relationship starts it is a good deal harder to scale back an existing commitment, particularly to allow the consultant to advise another company.

Besides defining a "Field," the contract will state the services you are to provide. You can control what is covered and what is not by defining the services specifically. If this issue is important to you, be mindful of open-ended commitments (e.g., "such services as the parties may mutually agree upon from time to time") and lists that are described as "examples" or that "include" the items identified.

> ### Example
>
> The draft contract states "Consulting services shall *include, without limitation,* assessing XYZ active compounds and advising about follow-up assays." An example of a more specific alternative is "Consulting services shall *consist of* assessing XYZ active compounds furnished by the Company for affinity to antigen, and advising the Company's scientists about follow-up affinity assays."

Also, it is at this point in the contract that it is appropriate to delineate preexisting obligations to the university.

> ### Examples
>
> Consider including this sentence in the section of the contract on services: "This Agreement is subject to the policies of [name of university] as they may exist from time to time, and anything in this Agreement to the contrary shall be void." Or "The Company acknowledges that the Consultant is a University employee and is subject to the University's policies, including policies concerning consulting, conflicts of interest, and intellectual property. The Company acknowledges and agrees that nothing in this Agreement shall affect the Consultant's obligations to the University, the Consultant's research on behalf of the University, or research collaborations in which the Consultant is a participant, and that this Agreement shall have no effect upon transfers (by way of license or otherwise) to third parties of materials or intellectual property developed in whole or in part by the Consultant as a University faculty member."[1]

Finally, if you are already consulting for another company in this same field, you will want to say that here, even if the new company knows about it and says it is not an issue. This subject can be very sensitive, and you would not likely want to omit it from the contract.

5 | Cash Compensation

C OMPENSATION IS OFTEN THE ISSUE OF most immediate interest to a prospective consultant. A contract should be clear about the elements of compensation to avoid future misunderstandings or disputes. Some specific issues relating to cash compensation are discussed below (stock and stock options are discussed in Chapter 6).

As mentioned elsewhere in this book, to the extent the discussion in this chapter refers to taxes and tax issues, we are providing a general summary of our understanding of tax laws. **You should not rely on any of our discussion as tax advice or otherwise. Please review Disclaimer on page xiii.**

A. Retainer

A retainer is a payment made to retain a consultant as an advisor to a company for a period of time. In many cases, a retainer is paid instead of an hourly or daily consulting fee (discussed below). If a retainer is to be paid in addition to a consulting fee, the retainer and the fee should be clearly distinguished from each other.

Retainers are generally paid on an annual basis in advance, not in arrears. It is obviously preferable to be paid in advance. For start-ups, a company might consider an allocation of founder's stock to serve as a retainer for consulting and/or SAB services, particularly if such shares vest over time and the company has not yet obtained substantial financing. (Vesting is discussed in more detail in Chapter 6.) In the start-up situation, a consultant might negotiate a relatively higher percentage of company stock or stock options than later in the company's growth cycle. It is also possible, with help of a tax advisor, to negotiate a retainer that will constitute deferred compensation consistent with tax laws so that payments, and hence taxes on such payments, will be deferred, to begin on or increase at a later date. Deferred compensation might be useful for those who wish to delay paying tax on current income, but these arrangements can be invalidated as improper tax deferral strategies if they do not satisfy regulatory requirements. So, as noted, expert tax advice should be sought as part of putting together any such arrangement.

A clause that automatically increases the amount of a retainer on a periodic basis over the life of a contract is advantageous by sparing the need to renegotiate and redocument.

> **Example**
>
> "The retainer will be increased by 5% on each anniversary of the Effective Date of this Agreement" or "The retainer will be increased by 10% on the third anniversary of the Effective Date of this Agreement."

B. Consulting Fee

A consulting fee is a payment made in exchange for consulting activity furnished over a unit of time. A consulting fee is standard, although as noted it might be excluded in lieu of a retainer or a grant of stock or stock options.

The consulting fee can be paid on a per-hour, per-day, per-month, quarterly, annual, or other basis. An annual consulting fee is equivalent to an annual retainer. If the fee is paid on a daily basis, the definition of a "day" should be clear. To avoid doubt, it might be useful to have the fee be paid "per day or portion of a day." Otherwise, a consultant might be paid for only a partial day at the company despite the fact that the company requested in advance that the consultant reserve an entire day. Similarly, consider whether you should be paid more for an extra-long day.

> **Example**
>
> "The Company shall pay the Consultant $2000 per day of consulting (to cover up to eight hours of Services, or any smaller increment thereof actually performed), plus an additional fee of $250 per hour for each additional hour of consulting services beyond eight hours furnished during any consulting day."

A common question is what is an appropriate rate for consulting services. There is no single answer, and many factors are relevant, such as: the importance of the consulting to the company's work; the relationship between the consultant and the company (e.g., is the consultant a founder? is the company building a platform around IP developed in the consultant's university laboratory?); the consultant's professional reputation, consulting experience, and compensation history; whether the consulting pertains to an invention licensed to the company of which the consultant is an inventor; whether SAB services are involved; and whether other compensation, such as equity, is involved. Ask around. The amounts are often higher, and sometimes significantly higher, for SAB membership or other special services; for a founder of a start-up; or for highly experienced and expert consultants.

Also bear in mind that generally the daily rate for a one-day or very short-term assignment can be higher than that paid under an ongoing long-term relationship.

Consulting fees can be paid in advance or, as is generally the case for hourly or daily fees, in arrears. A company might require an itemized invoice before paying. See if the company has a standard invoice form; using the form should save you preparation time and decrease the turnaround time for your check. In fact, many agreements require that you use the company's standard form. Beware of procrastination when it comes to billing—many corporate policies place hurdles on or even preclude payment of tardy invoices.

C. Travel Time

Travel time often is not covered. If there will be significant travel, consider whether time for travel should be compensated. If a company has a policy against paying for travel time, or otherwise refuses, some consultants raise their hourly consulting rate to offset the unrecovered cost of time on the road.

D. Expenses

The agreement usually will provide for reimbursement of reasonable out-of-pocket expenses. Save your receipts; as the case with invoices, some companies can be very particular about paying undocumented expenses.

There is not much more to say about this aspect of compensation, except to note that if you will be doing significant long-distance travel, a company might be willing to pay for business- or first-class airfare. Such a benefit might be another way to come to terms with not being paid for travel time.

Also, although it is unusual, some consulting contracts provide that the company will pay the consultant's legal expenses incurred in negotiating the consulting agreement. The more important you are to the company's mission, the better are the chances of getting this provision. A key founder of a start-up might be especially well positioned to negotiate this provision. Companies typically will report your consulting income annually on an IRS Form 1099. This information is furnished to you and the IRS for tax reporting purposes. Sometimes companies will mistakenly report reimbursed expenses as consulting income on the Form 1099, thus overstating your income and increasing the tax due. Be sure to review each Form 1099, and ask the company to correct it if the information is wrong.

E. Taxes on Retainers and Consulting Fees

Consulting compensation is usually treated as income earned outside the scope of employment. As noted in the preceding section, such income typically is reported to the IRS as "miscellaneous income" on IRS Form 1099 rather than as employment wages on IRS Form W-2. Companies are not required to withhold tax on cash retainers or consulting fees and ordinarily will not do so. Thus, the agreement will likely say that tax payments are solely your responsibility. (Compensation paid other than in cash, including stock options, is discussed in Chapter 6.) For this reason, you should consider having extra tax withheld from your university paycheck or making quarterly estimated tax payments to avoid a substantial tax bill or even a penalty. Quarterly estimated tax payments are usually straightforward to calculate. If you need advice, an accountant can help.

F. Deferral of Compensation Income through Retirement Savings

Even though you might participate in a retirement savings plan (such as a 401(k) or 403(b) plan) through your employer, the IRS permits you additionally to defer up to a certain amount of consulting and other compensation reported on Forms 1099 through other retirement savings vehicles. For example, as a self-employed consultant, you can contribute to a Simplified Employee Pension (SEP) account, which you can establish with a mutual fund or other investment management firm in the same way you can establish any other individual retirement account (IRA). In 2011, for example, you could have contributed up to 25% of your consulting income or $49,000, whichever is lower. In this way, you might be able to substantially reduce your current taxable income. A tax advisor can provide helpful advice.

6 | Shares, Stock Options, and Taxes

THE ISSUES RELATING TO EQUITY—SHARES of stock and stock options— can be complicated, and we encourage getting expert help in structuring this aspect of compensation. Some highlights follow. We do not address partnership or limited liability company (LLC) interests here. We also do not address stock options that are traded on a public market or other forms of rights in or interests relating to stock, such as stock warrants and stock appreciation rights (SARs).

A. Shares and Options Generally

Shares of stock represent a portion of a company's ownership, or capitalization. Shares can be issued in the company's common stock or in another series of stock, such as preferred stock. Consultants are usually compensated in common stock rather than preferred stock, although occasionally consultants are invited to purchase preferred stock alongside investors, as might especially be the case with start-ups.

Common Stock and Preferred Stock

Common stock represents an ownership interest ("equity") in a company and typically carries the right to vote on company matters, such as election of directors and the right to receive dividends. Holders of preferred stock typically receive the right to receive a fixed dividend that is payable before dividends on common stock and special voting rights on certain corporate matters, and sometimes also the right to convert the preferred shares into shares of common stock (called "convertible" preferred stock). "Convertible" preferred stock can be exchanged at a fixed rate for common stock, which can be valuable if the price of the common shares or the dividend paid on common shares (or both) makes the common shares more valuable. Companies can issue preferred stock in one or more series, each having its own set of rights in respect of dividends or voting on corporate matters.

One advantage of preferred stock is its higher priority in bankruptcy. Holders of common stock of a bankrupt or insolvent company generally cannot recover assets until claims of creditors, bondholders, and holders of preferred stock have been satisfied. Creditor and bondholder claims typically are senior to those of holders of preferred stock, and so whereas preferred stock stands in front of common stock, there must be leftover assets following payment of creditor and bondholder claims for preferred stock to receive its preference.

There are other advantages of preferred stock. It can provide for an established annual return. Thus, for example, a company could issue a series of preferred stock providing for an 8% dividend, and this payout would have to be made before any dividends may be paid to common stockholders. Also, to the extent the preferred dividend is unpaid when a company goes bankrupt, it must be paid before common stockholders get their dividend. In addition, as discussed in Section 6.J, preferred stock can be structured to offer holders special rights upon a "liquidation event," which may include a sale, an initial public offering (IPO), or other change of control of the company.

The issues relating to equity of all types—common and preferred; stock options; restricted shares (discussed below); stock warrants; stock appreciation rights, etc.—can be complicated. It is important to understand the terms underlying any equity you receive. Although there are useful resources to help with understanding general concepts relating to preferred stock[1] and other types of equity, it is often a sound idea to get legal and tax guidance before accepting any equity awards.

Definition of Public Company versus Private Company

Companies can be public or private.

For purposes of our discussion, a private company is one that has no shares offered via the public markets.

A public company is one that has issued shares via a registered public offering, such as an IPO approved by the U.S. Securities and Exchange Commission (SEC). Afterwards the registered shares can be traded freely via a broker.

However, a company might not register all of its outstanding shares in a public offering; for example, it might choose not to register shares held by insiders or founders. So, as part of acquiring options or shares of a company for which you will consult, you should find out whether it is public or private. If it is public, then you should determine whether your shares are registered and thus can be freely sold once any holding restrictions lapse. Usually shares are freely transferable if they are acquired by exercising stock options that were issued under a broad-based stock option plan registered with the SEC. If not, then the company should explain to you how and when you can sell your shares under exemptions from the public registration rules. If the company is private, then you should consider to what extent you can have your shares included in an eventual public offering. See discussion in "Registration Rights" in Section E below).

Restrictions on Ownership—Resale Restrictions (under Law) and Vesting Restrictions (Required by Companies)

Whether the company for which you will consult is public or private, it is likely that the company will issue your equity (be it in the form of options or shares) with restrictions. All shares with resale or other restrictions are called "restricted shares."

One common set of restrictions follows from a company's being private: Your opportunities to sell shares will be narrowly proscribed. These resale restrictions would be found in the documents you sign at the time of acquiring the shares, as well as on the share certificate itself. (The list of restrictions appearing on a stock certificate is called the "restrictive legend.") Founders' shares in a start-up generally have restrictions of this nature and so are considered "restricted shares." Usually a company will not allow restrictions to be removed from the shares until it is certain that the restrictions are no longer legally required. This change would occur after the shares have been registered with the SEC, for example, or if the shares are not registered with the SEC you are proposing to transfer the shares in a private transaction permitted by SEC rules. The company and your tax or legal advisors would be able to help you prepare the documents required to remove the restrictive legend in these circumstances.

The preceding paragraph describes restrictions imposed by the securities laws. What can be confusing, though, is that there can be a second kind of restriction on shares, and that is a restriction on ownership of shares imposed by the company rather than by law. This restriction is called "vesting." Typically companies will not give a consultant full ownership rights in shares or stock options immediately. As is the case with its key employees and directors, a company will want to prolong the process by which a consultant acquires full title to shares or options awarded to her. By linking ownership with length of service, the company has a valuable tool to entice management and key consultants to maintain their affiliation with the company. So, for example, a company might award you 10,000 shares of the company's stock on the two conditions that (1) your ownership of that stock will accrue at the rate of 2500 shares per year over four years, with the first 2500 shares becoming available one year from the date of the agreement, and (2) you continue to serve as a consultant as of each anniversary date. Under this "vesting" process, you will be "vested" in the first 2500 shares on the first anniversary of the agreement if you are then still consulting for the company, and at that time you will acquire legal title to those shares. You can later terminate your consulting arrangement with the company without affecting your title to those shares (assuming your stock acquisition documents do not impose other restrictions such as granting the company a right to buy them back if you quit), but you will lose the opportunity to acquire the remaining shares.

Shares that have not yet vested are also commonly called "restricted shares," so the term might be confusing, as it refers to shares with legal restrictions on resale and shares with contractual terms governing when they become vested. To avoid confusion, when we refer to "restricted shares" we are referring to shares of private companies, or to shares of public companies that have not been publicly registered; and when we refer to "unvested shares" we are discussing shares that have not yet vested (or, to use the language of the Internal Revenue Code, shares as to which there remains a "substantial risk of forfeiture"). Thus, "unvested shares" are restricted before they vest, but even after they vest they might still be "restricted shares" because they have not been registered for sale with the SEC or no exemption from registration is currently available for them.

Let us turn now to the tax impact of vesting of unvested shares. (Please note the caveats in Section 6.F.1 below.)

If shares are transferred without any vesting restrictions—in other words, ownership of the shares is transferred immediately without any holding period requirement—then they are taxed as ordinary compensation income at the time they are granted. If they are issued with a vesting requirement, they are taxed at the time they vest (i.e., at the time you acquire ownership) as ordinary compensation income; subsequent gains are taxed at the long-term capital gains rate if sold more than a year after the award date. The amount of income taxed is based on the fair market value of the shares at the time of the grant, for immediately vested shares, or at the time of vesting, for shares that are not vested at the time of the grant, less the price (if any) paid to acquire them. Treatment as "ordinary compensation income" means this income is taxed as ordinary income (i.e., not as capital gain) for federal tax purposes and (assuming you are not a company employee) also is subject to self-employment taxes covering Medicare and Social Security.

The holder of unvested (restricted) shares might find a tax advantage by filing a form with the IRS called a "Section 83(b) Election." So might the holder of unvested shares acquired upon exercise of a stock option. See "Section 83(b) Election" under Section 6.F.5 below. There might be penalties for shares or options sold at below fair market value (FMV). See Section 6.F.5 below for a discussion of Section 409A under the Internal Revenue Code.

Stock Options

A stock option is the contractual right to purchase a specified number of shares of a company's stock (usually common stock) at a predetermined price per share (called the "strike price"). Stock options, like shares, can vest over time. Once stock options are vested (and in some cases even before the options are vested—check the contract), you have the right to "exercise" the option, that is, to acquire the underlying shares in accordance with the terms of the stock option grant. (See Section 6.B below.) If the stock options are "in the money" or "above water," then the current value of the underlying shares exceeds the strike price. Your hope, naturally, is that when your option vests the current value will exceed your strike price. If not, the option is "out of the money" or "under water," and, depending how long the exercise period is, you can watch and wait for the current value to rise above the strike price.

> **Example**
>
> "Within ten days of the execution date of this Agreement, the Consultant will be granted nonstatutory stock options (the 'Options') to purchase 10,000 shares of the Company's common stock, par value $0.01 per share, at an exercise price per share equal to the fair market value thereof on the date of grant as set by the Company's Board of Directors, and will vest at the monthly rate of one twenty-fourth (1/24) of the shares subject to the Options, or 416.67 shares per month. The Options will be governed by the Company's 2011 Stock Plan attached as Exhibit A hereto and will be evidenced by the Company's standard form of stock purchase agreement in the form of Exhibit B hereto."

A stock option can be used as a singular or collective noun. In other words, one person might say that she has a "stock option to buy 10,000 shares," whereas another would say that she has "stock options for 10,000 shares."

In addition to price, the key element in acquiring shares or stock options is not the number of shares involved but the percentage that the shares represent in the company's overall capital structure. As a company issues additional shares, the percentage interest in any shares already issued will decline. This process is called "dilution." Dilution is discussed under "Antidilution Rights" in Section 6.D.

Stock options come in two varieties. Incentive stock options, or "ISOs," are not subject to U.S. federal tax upon the grant or exercise of the option but rather the underlying shares are taxed only after sale, and if retained throughout the requisite holding period they are taxed federally not as ordinary income (currently generally 25%–35% for faculty, excluding self-employment taxes) but at the lower long-term capital gains rate (currently 15%).[2] (ISOs lose their special tax benefits if sold before the holding period expires.) Although ISOs offer better tax treatment than the other type of stock options, which are called nonstatutory stock options, ISOs can be awarded only to company employees and do not offer the corporation the same tax advantages as do nonstatutory stock options. Also, only corporations can issue ISOs. At present, partnerships, LLCs, and other entities cannot. Although a company

could make you an employee for this purpose, many university policies we have seen would not permit you to accept outside employment.
We mention ISOs for information purposes. The rest of our discussion
addresses nonstatutory stock options only.

The majority of stock options awarded to consultants are nonstatutory
stock options, also known as nonqualified stock options or "NQSOs."
That is because ISOs, as noted above, can by law be issued only to
employees. NQSOs are taxed at the time of exercise (or at the time of
vesting, if they are exercised prior to their vesting) as "ordinary compensation income," and subsequent appreciation is taxed at the capital
gains rate upon sale if held for over a year.[3] For this reason, founders
and others who are granted options at a stage when the options have
little value might want to exercise them and pay the minimal conse-
quent taxes on ordinary income with the expectation or hope that the
shares will increase in value and that increase will be taxed at the lower
capital gains rate.

Stock options are usually issued under a stock option grant form that
defines the conditions under which they can be exercised and sometimes
how the underlying shares can be held or disposed of after issuance.
The option holder typically signs a short "stock option agreement" that
defines the unique terms of the stock option award. The stock option
agreement is typically tied to a stock option plan that defines general
rules regarding all stock options issued by the company pursuant to a
stock option agreement like yours. Often the consulting agreement will
reference the stock option grant and refer to the stock option agreement
for specific terms. (See the example above.)

Occasionally the stock option language will say that the grant
is subject to the approval of the Company's BOD or its stock option
committee. It is very unlikely that the BOD or committee will fail to
approve—because if they did you would quit—but if you want the obligation to be ironclad from the start then you should ask the company
to strike this condition. Usually the company will agree and obtain the
requisite approval before you and they execute the contract.

It is possible but less common for all the terms of your grant of shares
or stock options to be fully contained within the consulting agreement
or in a stock option agreement that is not tied to a stock option plan. In
any event, but particularly if this situation is the case, you should review

the relevant documentation to confirm that it describes the terms as expected and that all provisions are acceptable. A lawyer or accountant might be helpful in confirming that the terms include provisions you have not thought about and that the tax effect of the grant is what you are expecting. Of course, a lawyer or accountant's assistance is especially worth considering if the grant is of significant current or potential value.

Important stock option provisions include the vesting terms (see Section 6.B) and the restrictions on what you can do with the shares after they vest and you acquire title. Most such restrictions are relevant only if the shares are not yet publicly registered. There might be restrictions about how you can sell the shares, and you might be required to offer the shares back to the company before seeking an eligible buyer. You might also be required to sell the vested shares back to the company in the event you quit or are terminated under the consulting agreement. Finally, you might be required to vote your shares in a certain manner with respect to the election of directors or other corporate matters. You should read the documents to understand the rules affecting your shares.

Shares you acquire directly from a company, whether as restricted shares, unvested shares, or shares issued upon exercise of a stock option, are often issued under a "subscription agreement" or a "stock purchase agreement." The document can be very brief or, depending on the stage of the company, could include restrictions on when and how you may sell your shares. As is the case with stock options, the restricted stock purchase documents might limit your ability to transfer or vote the shares, particularly if the company is not public or the company has not publicly registered your particular shares.

Depending on the terms of the award, holders of restricted shares or unvested shares can receive dividends and vote on company affairs, although neither right would apply to unexercised shares underlying a stock option. Whereas these are advantages of holding restricted shares instead of an NQSO, the question of which is better for you depends on the specific facts and circumstances. The tax treatment can be different in different settings and dependent on the nature of the company's legal status (e.g., corporation, partnership, LLC, etc.) It is important to discuss these matters with a tax advisor if the company offers you alternatives.

B. Vesting

As noted, stock options and restricted shares (such as founders' shares) usually "vest" over time. For consultants, typically the options or shares will vest in equal monthly or quarterly portions over a three- or four-year period, and as already noted a standard condition is that at each vesting date the consulting arrangement must continue to be in place. Oftentimes vesting begins only after an initial interval has passed. For example, 25% of all options or shares might vest after one year and the rest might vest annually over the next three-year period. Obviously, the more accelerated the vesting schedule, the better for the consultant, and no vesting at all is ideal. If the company cannot agree to no vesting, try to get as many shares as possible to vest on the day the agreement is signed, with the remaining shares vesting over the shortest possible vesting period, with monthly vesting instead of quarterly or longer vesting periods. Monthly vesting is fairly rare outside the start-up context.

Vesting might be tied to one or more "milestone" events rather than to a periodic schedule like those described above. You might see such a provision with start-ups or other early-stage companies, but otherwise it is pretty uncommon. Milestones might include events such as the closing of a financing, the signing of a corporate partnership, the filing of an Investigational New Drug (IND) Application or New Drug Application (NDA) with the U.S. Food and Drug Administration, the issuance of a key patent, or the company's initial public offering. Milestones tied to your individual performance might raise questions at your university.

C. Acceleration Events

One important issue to consider is whether your stock options or restricted shares will undergo immediate accelerated vesting or a release from restriction under certain circumstances, such as if the company is acquired by another company, or you are fired without justification (without "cause"), or upon your death or disability.

This issue is potentially crucial and often overlooked. For example, a consultant might have worked hard to help sell the company, only to find that the new owner does not require his consulting services and as a consequence the unvested stock and options—possibly a substantial fraction of his anticipated net worth—are now void. Or the consultant dies before her options vest, and her family cannot exercise them.

To guard against the unexpected, the agreement can provide for immediate vesting of all unvested stock and options on the occurrence of an "acceleration event," and to include all appropriate events within the definition (discussed further below).

The first places to look for provisions concerning acceleration events are in the stock option plan and stock option agreement or, in the case of restricted shares, the relevant company plan and the share purchase agreement. If appropriate acceleration events are described and clearly would apply in your situation, then you might feel adequately covered. If not, it is up to you to ask the company to address accelerated vesting rights in your documents. Counsel or tax advice can be important in this context, because of what might be at stake and the nuances of the legal language. It is sometimes straightforward to negotiate for accelerated vesting when a company is just forming and the company itself sees little cost in agreeing to such a request. Later, when the cost of accelerated vesting might directly reduce the return to investors when the company is sold (the price paid for the company will in part be used to pay for the accelerated vesting), it might be difficult to persuade the company's BOD to grant such a request.

Example

"In the event (i) that the Consultant's consulting services are terminated by the Company without cause (as defined below), any unvested stock or stock options as of the date of such termination shall vest, (ii) of the death or disability of the Consultant (which renders the Consultant incapable of performing the services provided hereunder for more than 180 consecutive days), any unvested stock or stock options as of the date of death or disability shall vest, and (iii) of a change of control (as defined below) of the Company, any unvested stock or stock options as of the date of the closing or consummation of the change of control event shall vest immediately prior to consummation of the change of control. As used above, the term 'cause' means willful failure by the Consultant to perform the services required to be performed hereunder or material breach of this Agreement by the Consultant, which failure or breach continues for a period of 30 days after the Company provides the Consultant with written notice thereof. As used above, the term 'change of control' shall mean the occurrence of any of the following: (1) the dissolution or liquidation of the Company, (2) the sale of all or substantially all of the assets of the Company to an unrelated person or entity, (3) a merger, reorganization or consolidation in which the outstanding shares of the Company's common stock are converted into or exchanged for securities of the successor entity and the holders of the Company's outstanding voting power immediately prior to such transaction do not own at least a majority of the outstanding voting power of the successor entity immediately upon completion of such transaction, (4) the sale of all or a majority of the Company's common stock to an unrelated person or entity, (5) the registered offering of shares of any class of the Company's outstanding stock, or (6) any other transaction in which the holders of the Company's outstanding voting power immediately prior to such transaction do not own at least a majority of the outstanding voting power of the Company or a successor entity immediately upon completion of the transaction." Note: This example assumes that a majority of shares is appropriate to "control" the decisions and actions of the company. A smaller percentage, however, might have this power under the company's constituent documents, or a separate class of stock might hold it. It is important to understand the company's structure and shareholders' rights before negotiating this provision."

Consultants and corporate insiders will sometimes have "double-trigger" acceleration rights. Here the first trigger is change of control, and the second is involuntary termination. This means that both triggers must spring before the acceleration rights kick in. Double-trigger acceleration helps companies by making them more appealing as acquisition targets. With a double-trigger provision, key insiders and advisors cannot "cash in" upon sale but must continue to wait for vesting unless the new owner decides it no longer needs their services.[4]

Sometimes companies will ask advisors or insiders to give up (or "waive") their contractual right to acceleration if doing so will aid a financing or a sale or merger. The request might be a reasonable one if the transaction will raise the company's value so significantly that it makes more economic sense to waive rather than preserve the acceleration rights.

The issues relating to acceleration rights can be complicated and structuring them often requires legal or tax expertise.

D. Antidilution Rights

A company's issuance of new shares from time to time causes a decline in the percentage ownership of existing shareholders. Offsetting the decline in percentage interest, one hopes, is that the new shares are issued in exchange for cash or assets that increase the company's over-all value enough so that the existing shares do not diminish in value. Although that might be true when, for example, a VC firm invests in a start-up, oftentimes stock is issued without receipt by the company of any cash or marketable assets. Employee stock option plans are an example. In these cases the benefit to the company might be through intangibles (such as by motivating key employees), but there is no clear increase in the value of existing stock. Accordingly, the issuance of new shares can result in a "dilution" of the value as well as of the percentage ownership represented by one's existing share holdings in the company.

One way to avoid dilution is to have the right to increase your share ownership each time shares are issued so that your percentage owner-ship is preserved, i.e., to have "antidilution rights." Antidilution rights can be complete—e.g., a 5% holding will never be reduced below 5%—or might have a cutoff point, e.g., a 20% holding can be diluted down to 2% but cannot drop below 2%. Needless to say, antidilution rights can be very valuable.

Also needless to say, companies are reluctant to grant antidilution rights, even (albeit less so) to founders in a start-up situation, and estab-lished companies are unlikely ever to grant them. The company's reluc-tance arises from several factors. One is fairness if similar shareholders do not have antidilution rights. Another is that a smaller company often develops through VC or other institutional funding and these institutions will almost always demand termination of antidilution rights as a con-dition of investment. However, if you have sufficient importance to the continued viability of the company when a VC is ready to invest, you might be able either to maintain some or all of your antidilution rights or be compensated in some other manner in exchange for giving them up. Often enough, however, as noted above in the context of acceleration rights, the VC investment is so valuable to the company's prospects that it might make financial sense for the holder to waive her antidilution rights or to negotiate some payment or other benefit in lieu of antidilution rights.

If you obtain antidilution rights, the important question then becomes what constitutes a "dilutive event" that will trigger your right to receive additional shares. You will want to include any and all issuances of shares in the future, including shares issued to outside investors and shares issued upon exercise of stock options or other rights such as stock warrants. The company will want to keep the list of dilutive events narrow (excluding, for example, shares issued at fair market value) to make it easier to negotiate with future large investors.

Experience is key to maneuver effectively in this area, and a good business lawyer might be valuable in considering and negotiating these issues.

E. Registration Rights

As noted, a company might not publicly register all of its outstanding shares. In fact, often the company will publicly register only new shares that it wishes to sell to the public to generate capital for operations. Existing shareholders can publicly register their shares in what is called a "secondary offering," and typically (because of the costs and complexity) a shareholder will want the company to handle the secondary offering on his behalf. However, an IPO often will not include a secondary offering, and a company has no legal obligation to file a secondary offering unless it is contractually obligated to do so. And as noted, the costs and amount of effort might be too much for most consultants. Provisions relating to registration rights and secondary offerings are complicated and expert advice is definitely appropriate here.

There is a way for some shareholders to obtain the legal right to include their shares in a secondary offering. Well before a public offering—even at the time of formation—a company's management and key institutional shareholders will negotiate the terms upon which they might force the company to offer their shares, either through an otherwise planned public offering by the company (often called "piggyback rights"[5]) or a secondary offering of their shares alone ("demand registration rights"). These terms, called "registration rights," might also give one or more of these shareholders the right to force other parties to include their shares in a secondary offering (called "drag-along rights") when, for example, the underwriters believe the additional shares are needed to market the offering. Large later-stage investors such as VC firms might also seek registration rights to allow them to sell their shares through a secondary offering, and a VC firm might seek to be first in line ahead of company insiders.

If you are negotiating for restricted shares or stock options in a start-up or other privately held company, you might seek the same registration rights enjoyed by key shareholders. In general, the chances of a consultant's getting registration rights are low, but such rights have been granted. You have a better chance if you are a founder or an inventor of indispensable company-licensed technology.

The terms of registration rights typically are set out in a "registration rights agreement." These agreements almost always are long and very complicated and given the financial stakes involved should be reviewed carefully by an experienced legal advisor.

As an alternative, the securities rules (and specifically an SEC rule known as "Rule 144") might allow you to sell unregistered shares of a public company in the public markets subject to certain requirements, including how long you have held the shares and whether you are a corporate insider.[6] If this situation applies, you should seek help from the company or an experienced professional.

F. Some Thoughts about Taxes

1. Introduction: Caveats

The tax treatment of stock options and restricted shares is very complicated. Whole books are written about this subject, and blogs provide constant updates about the topic. And tax laws change frequently. We are far from experts in this area, so although we attempt to summarize some of the basics, you should not rely on this discussion. **We strongly recommend getting tax advice** when negotiating for, exercising, or selling stock options or restricted shares. As already noted, the discussion of stock options here relates only to NQSOs, not ISOs, and we do not address publicly traded options. Also, we do not examine issues relating to the federal alternative minimum tax (AMT). We do not discuss state tax issues or tax matters that might arise under foreign laws, whether based upon the location of the company, the citizenship of the taxpayer, or otherwise. Finally, as noted earlier, U.S. tax rules affecting partnership or LLC interests are not addressed here. If you are receiving a partnership or an LLC interest rather than shares or stock options, the rules are different and you should seek advice regarding their U.S. tax treatment.

2. Overview[7]

Shares of stock and stock options can be granted either outright or subject to restrictions, such as vesting. Either you are given the shares or options, or you pay the company for them. Once restricted shares vest, they may be sold depending on the specific restrictions applicable to the shares. Those restrictions appear in the document you signed to acquire them. This document usually is called a subscription agreement or a stock purchase agreement. Restrictions relating to transferability might also appear in the "restrictive legend" on the stock certificate. These topics are addressed in Section 6.B above.

As discussed earlier, to exercise a stock option, you pay the company a certain price (the "strike price"), after which the option is converted into shares of stock. You may sell those shares once they have vested, subject to the terms of the stock option documents and any restrictions that might be imposed under federal or state securities laws.

To the extent your consulting compensation includes a grant of shares (restricted or unrestricted) and the price (if any) you pay for the shares is lower than the stock's FMV at the relevant date (discussed below), you are deemed to receive "compensation income" for tax purposes to the extent of that difference. This "compensation income" is determined and is taxable at the time of grant if the shares are then vested or at the time of vesting if they vest later (and no Section 83(b) election was made; see below), and is taxed as ordinary income, not as capital gain. (Bear in mind that tax on compensation income might include FICA/Medicare or self-employment taxes.) In the case of a stock option, the compensation income is equal to the difference between the strike price and the stock's FMV at the time of exercise and is determined and is taxable upon exercise, rather than grant, of the option. (Note, although probably irrelevant here, that the rules are different for options that are publicly traded.) In any event, further appreciation in the value of that stock after you exercise the option or after vesting is taxed as capital gain at the time the stock is sold. The tax will be at the short-term capital gain rate if you sell the stock within one year after vesting or as long-term capital gain if held over a year after the vesting date.

As noted, the taxable event for federal tax purposes is triggered when your unvested shares vest or when you acquire shares underlying a vested option. The principle underlying the idea that the vesting of unvested shares is a taxable event is that you own the shares outright once vesting has occurred—or, as the IRS says, there is no longer a "substantial risk of forfeiture." For shares that vest, your income for tax purposes will be the difference between the purchase price you paid at issuance and the FMV of the shares on the date they vest (unless you filed a Section 83(b) election; see below). For exercised stock options, the income for tax purposes will be the difference between the strike price and the FMV of the options on the date of the exercise.

When receiving NQSOs, be sure your tax advisor is satisfied that the company set your exercise price at no less than FMV and that the company has properly made and evidenced this determination. Otherwise, under Section 409A of the Internal Revenue Code, you could have taxable income when the option vests (instead of at exercise) and owe in addition to the normal tax an additional 20% tax. The issue might not be a problem for a seasoned company with an established market for its shares. But determining the FMV can be particularly challenging for a start-up company or when valuing illiquid securities. Therefore, if you are in this situation, you and your tax advisor should confirm that the grant qualifies for "safe harbor" treatment under Section 409A, and although your determination of safe harbor status might still be challenged by the IRS, your actions could decrease the chances that the IRS could successfully impose additional taxes under Section 409A. With some exceptions, safe harbor treatment for illiquid securities of a start-up company is determined by the reasonableness of the pricing valuation as supported by a written report of a qualified appraiser or corporate valuation firm.[8]

Sometimes a company plan will allow you to pay via a promissory note (in essence, an I.O.U.) instead of cash. There are federal tax implications if the interest rate on the note is set below market. Again, this issue is one for your tax advisor.

Note that all examples of taxation here assume that the tax payable is the marginal rate of taxation. That is, all gain is taxed at the tax rate stated in the example.

Example (Vested NQSO)

On August 1, 2010, you were granted an NQSO for 10,000 shares, and the option vests annually in four equal annual increments in accordance with the company's stock option plan. The strike price is $5 per share, which is the FMV of the shares on August 1, 2010. On August 1, 2011 you exercise the option for the first 2500 shares and pay the company $12,500 for them or (if the plan permits) give the company a note for $12,500. The company issues you a stock certificate for 2500 shares on that date and informs you that the FMV is then $10 per share. You have received compensation income, and you will owe federal tax at the ordinary income rate based on the difference between the FMV of the 2500 shares on August 1, 2011 ($25,000) and the total strike price ($12,500).

Example (Restricted Shares)

On August 1, 2010, you purchased 10,000 restricted shares, and the restrictions lapse annually in four equal increments. The purchase price was $5 per share, and on that date you paid the company $50,000. The company issued you four stock certificates for 2500 each, and each has a "restrictive legend" setting out the duration of the restrictions on the shares covered by that certificate. One year later, on August 1, 2011, the restrictions on the first 2500 shares lapse, and you exchange the relevant share certificate for a new one with all the restrictions removed. You do not make a Section 83(b) election for any of these shares. You will owe federal tax for the 2011 tax year based on the difference between the purchase price and the FMV of the 2500 shares on August 1, 2011. If the FMV on August 1, 2011 is $10 per share, then you would owe federal tax for the 2011 tax year at the ordinary income rate based on the difference between the FMV of the 2500 shares on that date ($25,000) and the total purchase price you paid for those 2500 shares ($12,500).

In both of these cases, you generate "compensation income" for federal tax purposes, but the income is only "on paper," assuming you do not dispose of any of the shares during the current tax year. Note that the tax accrues immediately if there are no vesting restrictions or at the time vesting restrictions expire if there are vesting restrictions, regardless of whether you have sold the shares. In tax parlance such income is called "phantom income"—that is, taxable income with no actual cash received. If you lack other resources with which to pay the tax, you might have to sell the shares earlier than you prefer to pay your tax bill.

There are various reasons not to dispose of shares immediately. One obvious reason is if you think that their value will go up. Or if the shares are issued to you when the company is private, your stock option agreement or restricted share agreement likely will restrict how you may dispose of the shares as long as the company remains private.

If you hold the shares through the current tax year and the stock drops below the price you paid, generally you will still owe tax based on the formula described above (tax equals [FMV on the date of vesting or exercise, minus price paid] times [effective tax rate]), and the amount can be considerable. If you do not have cash available to pay the tax, this situation can be perilous, as many insiders learned when the "dot com" bubble burst. See the examples under "Tax Risks" below.

3. Tax Risks

We have described some situations that will generate phantom income, that is, taxable compensation income for which the taxpayer has received no cash. What this means, of course, is that if you are about to acquire shares either through the lapse of restrictions on restricted shares or through the exercise of a newly vested option, you should determine beforehand either that (a) if you cannot or will not sell the shares, you will have enough available cash to pay the tax, or (b) if the shares can be sold, you can immediately sell enough of the shares to cover the expected tax bill.

In these situations, the compensation income is not only taxable, but is taxable as ordinary income, not as a capital gain. That means it is taxed at a higher rate, and there might be FICA/Medicare or self-employment taxes owed.

This problem is not hypothetical. There are many stories of people who thought they were millionaires in dot-com or other cutting-edge companies who ended up owing the IRS a great deal of money, sometimes with financially disastrous results. Consider the following, based on a real-life example we heard about.

Background

Professor Jill Smith acquires stock options for 100,000 shares of a biotech start-up, Acme Corporation, at the time of its formation. Acme does great and eventually goes public. After her option vests, she exercises her option at a total strike price of $100,000 (or $1 a share), and she is happy to spend the $100,000 because the shares have a current market value of $50 each or $5 million in total. She thinks Acme stock will triple over the next two years, and so she holds onto all of the stock.

Case 1—No Sale

Let's assume Dr. Smith holds her shares without selling any of them during the tax year in which she acquired them. She will owe tax on the difference between the FMV on the date of exercise ($5 million) and her total acquisition price ($100,000). Thus, she would owe federal tax on $4.9 million of ordinary compensation income or, using today's marginal tax rate of 35%, she would have a federal tax bill of approximately $1.72 million.

Case 2—Sale for Loss

Let's assume that despite Dr. Smith's expectations, during the same tax year as vesting, Acme's stock begins to drop precipitously as the company's business encounters serious unexpected problems. Before the end of the year it is down 20-fold and her stock is worth only $250,000. The company's situation is so dire that Dr. Smith finally sells the shares before the end of the year for $250,000. Dr. Smith's tax basis in the shares was $5 million, the FMV when she acquired them (and on which her taxable compensation income was computed), so she will report a loss of $4.75 million. Because she held the shares for less than one year after the date of exercise, the loss is a short-term capital loss. As such, under current federal tax law, only $3000 ($1500 if married and filing separately) of this huge capital loss can be used to offset ordinary income in the current tax year. Thus, assuming she has no other deductions, Dr. Smith would owe the IRS a tax payment based on $4,897,000 of ordinary compensation income, or (similar to case 1) approximately $1.71 million, despite her realizing only $250,000 from the sale of her stock. She can apply the capital loss against other capital gains, now or in future tax years (such a loss is called a "loss carryforward"), and she can carry forward any unused capital loss to offset ordinary income in future years, but as noted no more than $3000 per year of capital losses can be applied against ordinary income in any year. See discussion of the $3000 capital loss rule in "Ordinary Income (Loss) versus Capital Gain (Loss)" below.

4. Ordinary Income (Loss) versus Capital Gain (Loss)

As already noted, federal tax laws in 2011 set two rate schedules for income—a higher rate for ordinary income or short-term capital gains, and a lower rate for long-term capital gains. Presently, for many faculty, ordinary income is taxed currently at 25%–35% (our examples assume a 35% rate); long-term capital gains on stock are taxed at 15%. (Again, compensation income is also subject to Medicare and Social Security taxes; if you are an employee, you pay half of these taxes and your employer pays the other half, but if you are not employed by the company, then you pay both halves as self-employment taxes.) These rates might increase soon based on recent tax proposals in Washington. You should check the IRS Web site or call your tax advisor to confirm ordinary income and capital gains rates, since these topics will continue to be actively discussed in the coming years.

As already noted, the taxable event triggered by exercising an NQSO or the vesting of restricted shares is treated as ordinary compensation income. Shares sold within one year after acquisition generate short-term capital gain (or loss), and shares sold after one year generate long-term capital gain (or loss). For this reason, it is advantageous under current federal tax law to wait at least a year to sell a profitable stock.

Under current tax law a short-term or long-term capital loss can be applied dollar-for-dollar against long-term capital gains in the current tax year, and any balance can be "carried forward" to apply against future long-term capital gains. However, as mentioned in case 2 above, only up to $3000 of capital losses (for a married couple filing jointly) can be applied against ordinary income in the current or any future tax year.

5. Section 83(b) Election for Restricted Stock

The Internal Revenue Code includes a provision that under certain circumstances, once restrictions on restricted stock lapse and the shares are vested, the shares may be sold within a year after the vesting date and nonetheless be taxed at the long-term capital gain rate, as though the stock had been held without restrictions for over a year. (This provision applies only to restricted shares and not NQSOs.) Under present federal law, the result of this situation is the shareholder might pay less tax. This Code provision, Section 83(b), governs federal taxation of property that is awarded as compensation for services rendered and that is subject to a "substantial risk of forfeiture"—for example, securities that are subject to a vesting requirement such as restricted stock. Section 83(b) allows you to accelerate the date of the "taxable event" to the date you received the restricted stock instead of the date on which the restrictions lapse. The Section 83(b) election also entitles you to value the restricted stock as of the date of acquisition rather than the date the restrictions lapse.

This difference can mean a much lower valuation. To be effective, under current law a Section 83(b) election must be filed with the IRS within 30 days after you first acquired the restricted stock. A copy of the election also must be included with your tax return for the year in which the election was made, and a copy sent to the company. You should seek assistance from your tax advisor in preparing and filing the election.

There are various implications of the Section 83(b) election, as discussed below.

First, if you do not file a Section 83(b) election you can find yourself owing more in taxes than you receive if and when you sell your shares. Specifically, absent a Section 83(b) election, the amount of ordinary income on which you are taxed as a result of receiving restricted shares is determined by the difference between the purchase price you paid for the restricted shares and the FMV of the shares at the time when the restrictions lapse. For growth companies such as start-up or other early-stage companies, the difference between the price paid to acquire restricted shares at formation and their value some years later when the shares vest can be significant. However, if the stock value plummets after your shares become unrestricted but before you sell the shares, you can owe taxes on a far greater amount of "income" than you receive from the sale (phantom income). Not only that, but if you sell shares that have just had their restrictions lapse to pay the tax triggered by the lapsing of the restrictions, you would pay tax on any additional appreciation at the time of the sale at the short-term capital gains rate, because the sale would be made within a year after the restrictions lapsed.

By making a Section 83(b) election, the obligation to pay ordinary income tax on the value of the stock is triggered not at the time the restrictions lapse but rather at the time the restricted shares are first issued to you: You are electing to accelerate the date of the taxable event. You pay tax based on the value of the shares as of the date the shares are sold or granted to you, even though the shares are unvested. You might have minimal taxable ordinary income at this point, either because you purchase the shares at the then-current FMV or because you are granted shares that are then of minimal value. In addition, by making the Section 83(b) election, you cause the holding period for long-term capital gain treatment to start on the issuance date, not the date on which the restriction lapses. If the shares have increased in value as of the time the restrictions lapse more than a year later and you then sell the shares, any increase in the value of the shares will be taxed as long-term capital gains, not short-term capital gains, because the one-year long-term capital gains holding period will have been satisfied.

There is a risk of making a Section 83(b) election: You will be paying ordinary income tax earlier than would otherwise be necessary or paying a tax that might never otherwise come due. If the share price stays constant, then your only loss is that you will have paid taxes earlier than necessary, and you will have lost the use of those funds for the interim period. If the share price falls, however, you will not only have paid taxes early and lost the use of funds, but you also will pay more ordinary income tax than you would have without the election, because you paid taxes based on the higher value of the stock at the time it was granted to you, rather than the lower price at the time the stock vests. And, if you forfeit the stock, you will lose both the purchase price you paid for the shares (which is usually quite a modest amount, or nothing at all) and the taxes you paid (although you might recover some of the loss of the purchase price of the shares as a capital loss).

Most companies are careful to issue founders' shares at FMV, especially after enactment of Section 409A (discussed earlier in Section 6.F.2). So in most cases the risk of filing a Section 83(b) election on sub-FMV shares should not be present. Nonetheless, prudence suggests that it is always sensible to request Section 409A safe harbor documentation as part of the Section 83(b) election process.

The following examples illustrate possible different results with and without a Section 83(b) election:

Case 3—No Section 83(b) Election

Year 1: Dr. Jack Jones bought 750,000 shares of restricted shares of a biotech start-up, Zebra Corporation, on the date of its formation, February 1, 2005. He purchased the shares at the FMV of $0.001 a share for a total of $750. The shares vest in equal installments annually over a three-year period beginning one year after issuance. He is a founder of and consultant to the company. If he stops performing consulting services for the company before the shares fully vest, the company will have the right to buy back the unvested shares for $0.001 per share.

Year 2: The first 250,000 shares vest on February 1, 2006. Assume Zebra's FMV on that date is $1 per share. Absent a Section 83(b) election, Dr. Jones is deemed to have federal taxable ordinary income of $249,250 for the 2006 tax year ($250,000 FMV as of the date of vesting minus $750 purchase price). Assume that he is in a 35% federal income tax bracket. Dr. Jones will owe approximately $87,000 in federal income tax in 2006 as a result of the vesting of the shares, plus related self-employment taxes. If the company is still private he may not be able to sell shares to pay the tax. If he does not have other financial resources, he could be in serious trouble with the IRS if he cannot sell shares during the tax year to satisfy his tax bill.

Year 3: The second tranche of 250,000 shares vests on February 1, 2007. Assume Zebra's FMV on that date is $4 per share. Again, assuming he made no Section 83(b) election, Dr. Jones is deemed to have $999,250 in federal taxable ordinary income for the 2007 tax year ($1,000,000 FMV as of the date of vesting minus $750 purchase price). At a tax rate of 35%, this income results in a federal income tax liability of approximately $350,000 for the 2007 tax year, plus related self-employment taxes. Again, if the company is still private he may not be able to sell shares to pay the tax and could be in serious trouble with the IRS if he lacks other resources with which to satisfy his tax bill.

Year 4: The third tranche of 250,000 shares vests on February 1, 2008. Assume Zebra's FMV on that date is still $4 per share. Again, assuming he made no Section 83(b) election, Dr. Jones is deemed to have federal taxable ordinary income of $999,250 ($1,000,000 FMV as of the date of vesting minus $750 purchase price). At a 35% tax rate, he owes another approximately $350,000 in federal tax for the 2008 tax year, plus related self-employment taxes.

Year 5: On January 5, 2009, Zebra goes public. No restrictions remain. (Note that oftentimes there is a lock-up period after an IPO during which shares held by company employees and consultants cannot be sold; for simplicity, we will assume there is no lock-up period in the case of Zebra, and Dr. Jones is able to sell his shares immediately.) The share price is $6. Dr. Jones sells all of his 750,000 shares. Ignoring for simplicity Dr. Jones's purchase price of $0.001 per share, his after-tax gain on the first 250,000 shares is $5 per share and on the remaining 500,000 shares is $2 per share, so his total taxable gain is $2,250,000. He owes federal tax at the 15% long-term capital gain rate on $1,750,000 (the total gain on the first two tranches, both of which qualify as long-term capital gain because he held them for more than one year) and at his 35% ordinary income tax rate on the $500,000 gain for the last tranche, which he held for less than one year. The total tax would equal $437,500 ($262,500 plus $175,000). All in all, Dr. Jones sold his stock, for which he paid $750, for $4,500,000 and has incurred federal tax over four years of approximately $1,224,500 ($87,000 plus $350,000 plus $350,000 plus $437,500), plus self-employment taxes.

Case 4—Section 83(b) Election

(Same facts as case 3, except that a Section 83(b) is properly and timely filed)
Dr. Jack Jones bought 750,000 shares of restricted shares of a biotech start-up, Zebra Corporation, on the day of its formation, February 1, 2005. He paid $750 (or $0.001 a share, the FMV) for the shares, which vest in equal installments annually over a three-year period. He is a founder and consultant. Let's assume that Dr. Jones timely filed a valid Section 83(b) election covering all 750,000 restricted shares.

Year 1: Dr. Jones will owe no federal tax at this point, since he paid the FMV for the shares and thus there is no "gain" that would trigger a federal tax obligation. He will owe no tax when his shares vest but rather will owe tax only when he sells the shares, and any appreciation in the value of the shares will be taxed as capital gain, not ordinary income.

Years 2–4: No taxable events occur.

Year 5: On January 5, 2009, Zebra goes public. No restrictions remain. The share price is $6. Dr. Jones sells all of his 750,000 shares for a gain of $4,499,250 ([750,000 × $6] – $750), which is taxed at the 15% rate applicable to long-term capital gains, because the one-year holding period has run or is deemed to have run. Dr. Jones's tax bill is approximately $675,000.

Recap: In this example, Dr. Jones's total federal tax bill after a properly filed Section 83(b) election is $675,000 instead of $1,224,500 without one. In this case, the Section 83(b) election saved him roughly $549,500 in federal tax. And importantly, it saved him from owing tax prior to generating cash from the sale of his stock. If not for the Section 83(b) election, Dr. Jones would have owed nearly $800,000 in years 2–4 and had no corresponding proceeds to pay it.

Does the Section 83(b) Election Make Sense for You?

Given these examples, it should be clear that properly filing a Section 83(b) election makes enormous sense in circumstances like those in cases 3–4, specifically, if the company is an early-stage or "growth" company and the amount of tax that would have to be paid early as a result of filing the election is low (in this case, zero). If the amount of tax paid early is nominal, there is very little at risk.

Would it ever make sense not to file a Section 83(b) election? In short, yes. For a mature company that has stock of some value at the time of the stock purchase or grant, or even a less-mature company that has a stock of value, you would run the risk of paying taxes on shares that might fall significantly below the value that you paid taxes on, or that you may forfeit. Or you might not want to file a Section 83(b) election if the risk of forfeiture of the stock is high—for example, if there is a substantial likelihood that you might stop performing services for the company before the stock vests, or you infer that changes in the company could lead to your early termination and your contract does not protect you (with accelerated vesting) in such event. In these cases, the company might have the right to repurchase the stock at your cost (or you may forfeit the stock outright, if you paid nothing for it), and you will have paid tax on stock you do not own and hence can never sell.

Again, we are not giving tax advice. In every case you should seek independent advice if you are unsure of the tax laws.

Should You Make a Section 83(b) Election after You Exercise a Nonqualified Stock Option (an NQSO)?

If you exercise an NQSO and the shares you acquire are vested, you no longer face a "substantial risk of forfeiture" and so the exercise immediately generates a taxable event. However, if the shares that you acquire when you exercise an NQSO are not vested, you then will have shares that are restricted until they vest. You could in principle file a Section 83(b) election promptly following the purchase of shares upon exercise of an unvested NQSO. Does it make sense to do so? The answer is much like the answer discussed above—it depends upon weighing the risk of losing the cash spent at the time of exercise (in this case the sum of the exercise price and the tax paid if the exercise price is less than the FMV) against the benefit of paying taxes later based on the value of the shares at the time of vesting. If the cost of the exercise is minimal and the exercise price is close to the FMV, you would be putting very little cash at risk and properly filing a Section 83(b) election might make sense. On the other hand, if the strike price is high and/or there is a substantial difference between the strike price and the FMV at the time of exercise causing you to owe considerable taxes immediately, filing a Section 83(b) election might prove costly.

Last Word on the Section 83(b) Election

The issues concerning whether or not to file a Section 83(b) election can be complicated, and once again we encourage you to have an appropriate discussion with your tax advisor at the planning stage.

6. Gifts and Estate Planning

If you are contemplating giving stock as a gift or leaving stock or the proceeds from stock in your estate (for example, to your children or other family members), additional tax considerations become relevant. Gift and estate taxes can be substantial if the amounts involved are major. There are a variety of legal vehicles that can reduce or even eliminate such taxes. For example, if founders' shares are assigned at the time a company is founded not to the founder but rather to an appropriately defined irrevocable trust with the founders' children as beneficiaries, the gift to the trust can be minimal and estate taxes can be avoided. Similarly, Roth IRAs and Grantor Retained Annuity Trusts (GRATs) can be used to postpone and reduce gift and estate taxes. The alternatives are many and can be complicated, and again an appropriate discussion with your tax advisor is important.

G. Fair Market Value

As has been noted, tax is paid on the difference between fair market value (FMV) and the strike price (for options) or purchase price (for shares). A public company's stock price is listed in the daily newspaper or online, and so it is easy to determine the FMV. But how does a private company, and in particular a start-up, define the FMV of its shares? Usually in these situations the company's BOD will determine the FMV. Precisely how they do so can be complicated. Given changes in tax laws that affect private company stock valuation, private companies often look to independent appraisers to value their stock. Many companies traditionally had set the price of common stock as a fraction of the price of the most recent round of preferred stock. However, this practice is becoming less standard. No matter how the FMV is determined, the IRS might challenge the value, and, if successful, the tax consequences could increase. See the discussion of Section 409A of the Internal Revenue Code in Section 6.F above.

H. Royalty Interest

As an alternative to an equity interest, a consultant might inquire about a royalty interest on an invention or platform. If the consultant's work will be integral to a product development, then getting a percentage of royalties on all products sold from the platform might be more valuable, especially as a royalty interest is by definition protected against dilution. The downside of this approach includes not only the less predictable nature of a future royalty but the length of time it might take before seeing any income. There are also matters to consider in terms of how the royalties are structured and accounted for; for example, if the interest is in a "net" royalty, then it is important to understand what items will be netted from royalties before payout and how they would be accounted for and audited. Although we feel we should mention this possibility, we are not aware of any consultant who has negotiated for a royalty interest.

I. Interest in a VC Firm

As noted, VC firms often retain experts to help them evaluate the IP platforms, science and business plans of potential investments. Those experts might prefer taking an interest in the VC firm rather than or in addition to shares or options in the firm's portfolio companies. An interest in the VC firm (usually called a "carry" or "carried interest" in the firm's profits) can be more lucrative, since it encompasses all investments, not just the ones reviewed by the expert (which might not be as successful as the others or might not succeed at all). Carried interest can be a complicated legal term because, for example, the carried interest could apply only to deals completed while the consultant is active with the firm. Also, the tax treatment of carried interests is under evaluation in Washington. This area, of course, is another for expert advice.

J. Liquidation Preferences and Participating and Nonparticipating Preferred Stock

Founders might be able to receive or purchase preferred shares of a start-up company. Preferred shares generally have certain rights not available to common shares, including liquidation preferences and participation rights. In general, nonfounder consultants are not offered the opportunity to obtain preferred shares. Nonetheless, an understanding of preferred shares is important to consultants who receive common stock, since the rights of common stockholders to share in proceeds from the company's sale or other significant event can be greatly affected by how the company's preferred share rights are structured.

We preface our discussion of liquidation preferences and participating preferred and nonparticipating preferred stock by cautioning that these terms are complicated and are very much the creatures of contract. Said differently, these terms can vary in definition from company to company, as discussed below. For example, a liquidation event, which involves the converting of stock to "liquid" assets, might include some or all of a sale, lease, license, or merger of the assets of the business or an IPO. The financial consequences of a liquidation event will vary depending on the liquidation amount and the relative rights of the holders of all classes of stock, which are all a matter of negotiation, with different outcomes for different companies. An inexperienced consultant likely will require expert guidance to understand the implications of specific liquidation preferences and participation rights.

Generally, upon a liquidation event (as that term is defined in the company's preferred stock purchase agreement or other governing contract), a holder of nonparticipating preferred stock will receive the liquidation value of her preferred shares (plus accrued and unpaid dividends, if any) or, if the preferred shares are convertible into common stock, the value of the shares as if converted to common stock immediately prior to the liquidation event, whichever is greater. The following is an example involving nonparticipating preferred stock. We assume for simplicity the common shareholder is fully vested and holds shares rather than options.

Case A

Bravo Biopharmaceuticals has issued 5,000,000 shares of Series A convertible nonparticipating preferred stock, each with a $10 preference upon liquidation because the holders of the preferred stock had invested $50 million. Bravo has also issued 5,000,000 shares of common stock. Dr. Smith is granted 100,000 shares of common stock, constituting 2% of the common stock and 1% of the total issued stock of the company. Bravo agrees to sell all of its assets for $50 million, and the asset sale constitutes a liquidation event under the Series A purchase contract. There are no accrued and unpaid dividends on the preferred shares. The Series A holders will take their liquidation preference distribution and receive all $50 million. Nothing will be distributed to the common shareholders, including Dr. Smith. On the other hand, if Bravo sold its assets for $200 million, the Series A holders would not take their preferred distribution, and instead their preferred stock would be treated "as converted" to common stock. The $200 million would be distributed among the 10 million shares, and Dr. Smith would receive 1% of the total sale price, or $2 million. By contrast, if Bravo sold its assets for $80 million, the Series A holders would take their preferred distribution (because converting to common stock would give them half of all proceeds, or only $40 million), leaving $30 million to be distributed among the holders of common stock, and Dr. Smith as a holder of 2% of the common stock would receive $600,000.

On the other hand, a holder of participating preferred stock will receive upon a liquidation event the sum of the value of her preferred shares upon liquidation *plus* the value of her preferred shares as if they were converted to common stock immediately prior to the liquidation event. Consider the following example.

Case B

Same facts as in case A above, except the Series A convertible preferred shares are participating. Again, Dr. Smith would receive nothing after a $50 million sale. After a $200 million sale, the Series A holders would receive $50 million from their liquidation preference and then receive half of the $150 million that remained. Of the $75 million remaining for the original holders of common shares, Dr. Smith would receive 2%, or $1.5 million. If Bravo sold its assets for $80 million, the Series A holders would take their liquidation preference of $50 million plus half of the remaining $30 million. Dr. Smith, as a holder of 2% of the common stock, would receive $300,000 of the $15 million remaining for common shareholders.

Thus, participating preferred stock gives the holder a guarantee of preferentially receiving funds from a liquidation event plus the opportunity to share in a greater gain. The holders of common stock gain less if there are also holders of participating preferred stock. We note that the advantages given to holders of preferred stock are generally in exchange for the greater financial risk taken by such shareholders, who are often venture capital firms or other institutional investors.

Sometimes, the liquidation preference associated with preferred stock involves a multiple, i.e., the holders of the preferred stock receive prior to a distribution to common shareholders not the amount correspondent to their investment but rather a multiple of that amount. Here is an example.

Case C

Same facts as in case B above, except the Series A convertible participating preferred shares have a 2× liquidation preference, i.e., they have a $100 million liquidation preference in exchange for their investment of $50 million. Again, Dr. Smith would receive nothing after a $50 million sale. After a $200 million sale, the Series A holders would receive their $100 million liquidation preference and then receive half of the $100 million that remained. Of the $50 million for the original holders of common shares, Dr. Smith would receive 2%, or $1 million. If Bravo sold its assets for $80 million, the Series A holders would take as their liquidation preference all $80 million, and Dr. Smith and other common shareholders would receive nothing.

Often, there are limits (known as "caps") placed on the extent of the liquidation preference of participating preferred stock, eliminating the participation in the event of a sufficiently high sale price.

Case D

Same facts as in case B above involving Series A convertible participating preferred shares (without the 2× multiple for the liquidation preferences), except the preferred shares are participating preferred shares with a "2× cap," meaning that the preferred shareholders' return will be "capped" at 2× the cost of their shares (i.e., 2× the value of their preferred liquidation preference), unless they convert their shares to common shares. After a $50 million sale, Dr. Smith would receive nothing. After a $200 million sale, the preferred holders would have a $50 million liquidation preference and could "participate" in the remaining $150 million up to the point that their total receipts do not exceed two times their liquidation preference, or $100 million. Thus, they would get $50 million of the $150 million, for a total of $100 million. Alternatively, they could convert their shares to common shares, but the result would be the same: As holders of half the common shares, they would receive half of $200 million, or $100 million. In either case, Dr. Smith, with 1% of the total or 2% of the common shares, would receive $2 million. If Bravo sold its assets for $80 million, the preferred holders would not convert and would receive $65 million ($50 million + half the remaining $30 million), since converting would give them only half of $80 million, i.e., $40 million. The outcome for Dr. Smith would be the same as in the $80 million example in case B: He would receive $300,000, i.e., 2% of the $15 million available for distribution to the original common stockholders.

The examples are depicted graphically in Figure 6.1, which compares the returns to holders of preferred and common shares in the four cases described: case A, nonparticipating preferred shares with a liquidation preference of an amount correspondent to the investment made by the holders; case B, participating preferred shares with a liquidation preference of an amount correspondent to the investment made by the holders; case C, participating preferred shares with a liquidation preference of a multiple of 2× the amount of the investment made by the holders; and case D, participating preferred shares with a liquidation preference of an amount correspondent to the investment made by the holders and a cap on the participation of 2× the amount of the investment. It should be clear from these graphs that the return to a common shareholder can

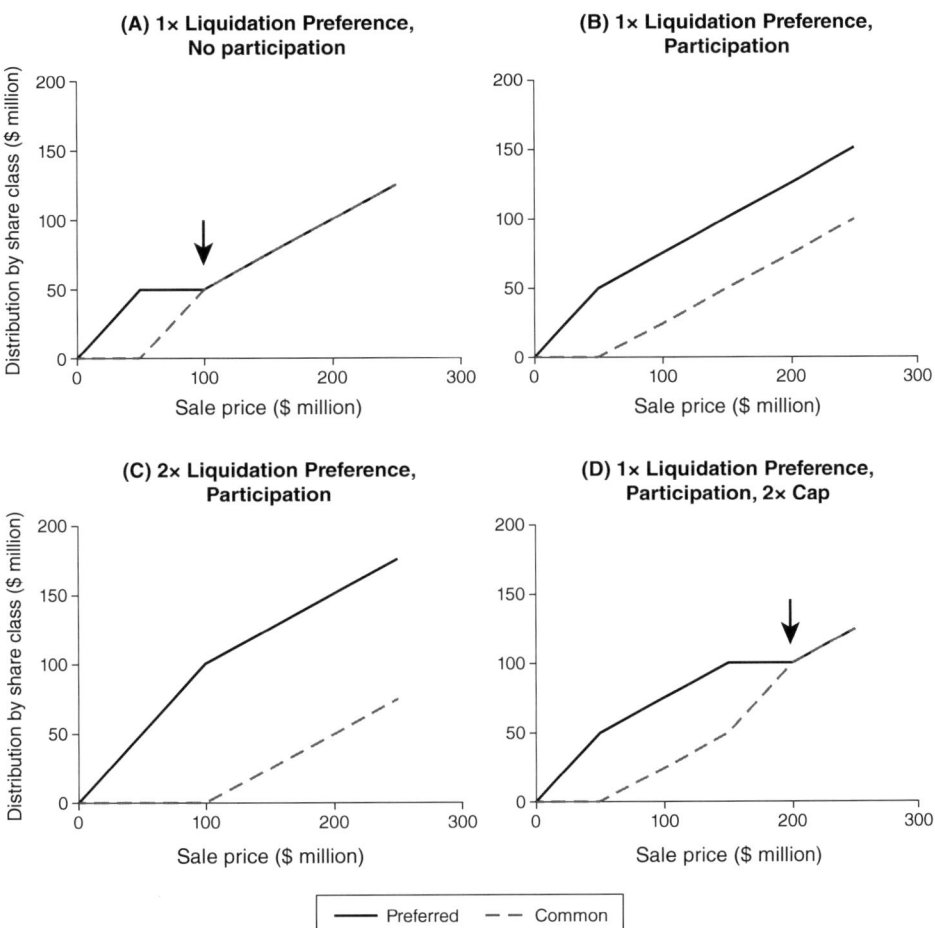

Figure 6.1
Examples of distributions ("Distributions by share class") after the sale of a company at dif-
fering prices ("Sale price") to preferred (black line) and common (dashed line) shareholders
given different cases of liquidation preferences and with or without participation. The four
cases are described in the text. The arrow indicates the point at which preferred shareholders
would convert their shares to common shares. (A) Nonparticipating preferred shares with a
liquidation preference of an amount correspondent to the investment made by the holders.
(B) Participating preferred shares with a liquidation preference of an amount correspondent
to the investment made by the holders. (C) Participating preferred shares with a liquidation preference of a multiple of 2× the amount of the investment made by the holders. (D)
Participating preferred shares with a liquidation preference of an amount correspondent to
the investment made by the holders and a cap on the participation of 2× the amount of the
investment. Note that for the general case described in the text in the absence of any liquidation preferences and without participation, preferred and common shareholders would own
equal numbers of shares and share equally in any proceeds.

differ widely in the four cases and is highly dependent upon the specifics, including the price at which the company is sold. Worth particular note is that in cases A and D there are sale prices at which an increased price results in an increased return to the common shareholders but not to the preferred shareholders ($50–$100 million and $150–$200 million, respectively), which means that the interests of these two groups are not aligned within these ranges.

These examples are relatively straightforward, but corporate finance is often more complicated. Companies often issue multiple series of preferred shares or other securities, such as convertible bonds, warrants, or stock appreciation rights, and each will have its own terms and economic incentives. There can also be new classes issued at later times that have more favorable terms that weaken (or "dilute") the value of existing shares. Investments can be tranched, i.e., staged over time, and a major "investment" can be made just prior to the company's sale or other significant event, greatly reducing the return to common shareholders if the new preferred shares are participating with a liquidation preference of substantially more than 1×. Complex spreadsheets are often needed to model the outcome for each class in a variety of liquidation scenarios. Such analyses can be challenging and are often best reserved for experts. Whatever the situation, each shareholder should feel comfortable understanding where her class of shares stands in the financial order as well as how it might be affected by future events.

K. Conclusion

We hope we have convinced you that there are complicated contractual and tax issues relating to the acquisition of stock options or restricted shares. Our take-home message is simple: Be sure you understand the relevant issues and seek expert advice in advance when appropriate.

7 | Confidentiality Obligations

ARLIER (IN SECTIONS 3.H AND 3.I) we spoke about confidentiality obligations owed to companies under confidential disclosure agreements, or "CDAs." Needless to say, confidentiality is a key element in consulting agreements, too.

As demonstrated by cases noted in the Introduction and Chapter 3, the stakes can be high when an academic scientist divulges confidential information belonging to a company for which he or she consults. This chapter addresses issues relating to confidentiality.

There are two directions in which confidential information can be disclosed in the course of a consulting relationship. One is the disclosure to you of the company's trade secrets or of proprietary information furnished to the company by its clients, collaborators, or others and then transmitted to you. It is this direction of information flow that the company cares about and will certainly address in the contract. The other is disclosure by you to the company of confidential information you possess about which the contract might be silent. We address the latter situation first.

A. Confidential Information in Your Possession

It is in your interest to state in the agreement that you have no obligation to share confidential information you might possess, whether such information is from your laboratory or communicated to you by others with the expectation of confidentiality.

However, as noted, consulting agreements drafted by companies often do not address your obligations of confidentiality, or to the extent that they do might be opposite in intent and actually include a broad obligation to inform the company of developments in the Field, regardless of your obligations to the source.

Example

"In addition to his other obligations, the Consultant shall discuss with the Company any information he may obtain regarding the Field, including without limitation new inventions, discoveries, and publications."

To protect yourself and your university research, and to control company expectations, consider building in standard caveats regarding any such disclosure.

Example

"In addition to his other obligations, the Consultant shall discuss with the Company any information he may obtain regarding the Field, including without limitation new inventions, discoveries, and publications, excluding any nonpublic information concerning his University research or other information to which he may owe an obligation of confidentiality." Or "The Consultant may disclose to the Company any information that the Consultant would normally freely disclose to other members of the scientific community at large, whether by publication, by presentation at seminars, or in informal scientific discussions. However, the Consultant shall not disclose to the Company information that (i) is proprietary to the University and (ii) is not generally available to the public, except through formal technology transfer procedures."[1]

A general exclusion for university and third-party information is always sensible.

Example

"Notwithstanding anything in this Agreement to the contrary, the Consultant shall not disclose to the Company any information that is confidential to the University or any third party."

As we noted in the Introduction and elsewhere throughout this book, the stakes are higher for those involved with clinical trials or other research of potentially imminent commercial importance, especially if they are consulting for a firm that might use the information for stock trading. Thus, clinicians and others in this category might wish to be even more explicit:

> *Example*
>
> "Notwithstanding anything in this Agreement to the contrary, the Consultant shall not disclose to the Company any information that is confidential to the University or any third party, including without limitation any unpublished results of a clinical trial, or any patient information or data, whether or not protected by the Health Insurance Portability and Accountability Act (HIPAA)."

As already noted, contractual protections such as these might be worthless if you choose not to follow them. The risk to your university research or to your interactions with collaborators or other companies might be substantial.

The stakes can be high for breaching a confidence. You do not want to be contractually obligated to do so.

B. Company Confidential Information

Companies have a legitimate interest in protecting their secrets. Therefore, consulting agreements typically require consultants to keep company information confidential. If the pharmaceutical companies that sponsored the clinical trials mentioned in *The Seattle Times* case (see Chapter 1) obtained a standard confidentiality covenant from the clinicians involved, the covenant would be a basis for any lawsuits that those companies might file against the clinicians. This provision is important for companies, and for this reason companies are very careful about how the confidentiality provisions are written into consulting contracts. It is important that you as a consultant be comfortable with what is covered.

1. Obligation to Keep Confidential

The company might draft the confidentiality provision very broadly in terms of scope and duration.

> *Example (Favorable to Company)*
>
> "The Consultant shall hold in trust and confidence, and shall not disclose, all Confidential Information, for the duration of this Agreement and thereafter."

We have two comments about this example. First, it has never been clear to us what the word "trust" is meant to cover in the example. As a legal term, "trust" implies a fiduciary duty, which is the highest level of responsibility recognized in business law. However, consulting is not usually considered to be a fiduciary role, and there is no reason for a consultant to assume a duty that might imply a fiduciary obligation. It is our experience that companies will typically agree to strike the word "trust" and simply require you to hold the Confidential Information in "confidence."

Second, the duration of the undertaking in the example above is unlimited, which means that the company can sue you for a breach of contract well after the relationship has ended. Because it is commonly understood, especially in cutting-edge areas of technology, that a new idea can quickly become outdated, our experience has been that companies will agree to limit the confidentiality obligation to the life of the contract and a set amount of time thereafter (a "tail period"). The range of the tail period is typically two to five years. Sometimes, although it is less common, the company will agree to limit the confidentiality obligation to the life of the contract only, with no tail period. For a consultant, the shorter the time period, the better.

Occasionally a contract will have a tail period triggered by the date of disclosure rather than the date of termination. This situation is advantageous by shortening the length of a confidentiality obligation. For example, assume you sign a five-year contract with a five-year tail period following termination. Further assume that on the first day of your contract you are told about a confidential "XYZ Analysis" and then you hear nothing about it again; or even consider that they tell you a week later that the XYZ Analysis was abandoned. Under the agreement, you would have to keep the XYZ Analysis confidential for ten years (five plus five). However, if the tail period were to run from first disclosure, the tail period would be only five years.

Example

"The SAB Member hereby agrees, for a period of five years following her first receipt of such Confidential Information, not to disclose or make use of, or allow others to use, any Confidential Information, except to the Company's employees and representatives, without the Company's prior written consent."

An even better provision would be to add a postterm cutoff. In other words, if you learn something on the last day of a five-year contract, your confidentiality obligation will extend past termination for some period short of five more years.

> **Example**
>
> "The SAB Member hereby agrees, for a period of five years following his receipt of such Confidential Information, not to disclose or make use of, or allow others to use, any Confidential Information, except to the Company's employees and representatives, without the Company's prior written consent; *provided, however*, that in any event all such obligations will expire no later than three years after the termination or expiration of this Agreement."

2. Definition of Confidential Information

It is important to consider how the company defines "Confidential Information" (which sometimes is called "Proprietary Information" or "Company Information"). If written poorly, this definition could be too broad and create doubts later about what is covered and what you can and cannot disclose.

Here is a pro-company provision:

> **Example (Favorable to Company)**
>
> "As used in this Agreement, 'Confidential Information' includes, without limitation, all trade secrets and confidential or proprietary information (and any tangible representation thereof) owned, possessed, or used by the Company, learned of by Consultant or developed by Consultant in connection with the consulting services hereunder, including (i) all Consultant Inventions [i.e., inventions made by Consultant in course of providing consulting services to company], marketing plans, business strategies, financial information, forecasts, personnel information, and customer lists of the Company, (ii) any sequence or other chemical or biological information or information concerning any other biological materials furnished by the Company to Consultant, and (iii) all information of third parties that the Company has an obligation to keep confidential."

Note the problems with this definition of "Confidential Information":

- The company is not required to indicate whether information it is giving you is confidential or to memorialize what is confidential in writing. Thus, if the company were later to accuse you of disclosing or misusing Confidential Information that you have no recollection of ever seeing, you would be in the position of having to prove that you never saw it. It is harder to prove the negative than to show that the confidential documentation you received did not include the relevant item. Thus, you should consider limiting the definition of Confidential Information to items that have been marked as confidential or, if conveyed verbally or electronically, are subsequently set forth in written form and marked as confidential. This request is usually acceptable. In the case the company will not agree to such terms, some consultants will sign anyway with the expectation that they will keep copious notes of all information the company shares with them and that these notes will be their protection if a problem were ever to arise. Although a good idea in the circumstances, we feel that it is better to require the company to put confidential information in writing. The note-taking approach has inherent challenges—for example, it is a common behavior of human beings over time to become less vigilant in any endeavor, including taking notes. Moreover, insofar as notes are incomplete by nature, there is a risk that the notes will be inadequate to disprove receipt of information not identified in the notes. In addition, it is unclear to us how far a note-taker should go to ensure the probative value of the notes in court; for example, should each set of notes be notarized promptly after they are completed?

- The definition covers third-party information shared with the company in confidence, but as written the company is not required to tell you that it is confidential third-party information. The problem here is the same as the one noted in the preceding bullet point—you could later be accused of disclosing confidential information without your ever having been informed that the information was confidential. The problem can be ameliorated by requiring all third-party confidential information to be identified in writing. Again, this solution is often acceptable.

- The inclusion of "Consultant Inventions" within the definition means that you cannot publish or otherwise disclose your inventions for the company during the confidentiality period. In other words, unless you have the company's permission, you cannot publish a discovery you make for the company until the confidentiality period ends. Furthermore, if "Consultant Inventions" is defined broadly, the company might assert rights in related work in which you are engaged outside the consulting agreement or after it is over. This issue underscores the importance of defining clearly what work will belong to the company and what will remain separate, for example, at the university. This topic is discussed further in Section 2.D and Chapter 8. For this reason, at least some universities and research institutions require a specific exclusion of university IP from the definition of Consultant Inventions, and such an exclusion is usually acceptable.

Here is an example of a definition more favorable for the consultant:

Example

"As used in this Agreement, 'Confidential Information' shall comprise all trade secrets and confidential or proprietary information (and any tangible representation thereof) owned, possessed, or used by the Company or learned of by Consultant in connection with the consulting services hereunder, in each case to the extent that each such item is marked by the Company prominently as 'confidential' upon first presentation to Consultant, or if first presented to Consultant other than in writing, is reduced to writing and marked by the Company prominently as 'confidential' and presented to Consultant within ten days after first disclosure, including Company Inventions." (This last phrase might be acceptable if the term "Company Inventions" is appropriately defined. See Section 8.A.)

3. Exceptions to the Definition of Confidential Information

The agreement usually will offer appropriate exceptions to the definition. A typical list of exclusions follows:

Example

"The term 'Confidential Information' hereunder shall not include information that the Consultant can establish by competent written evidence that (i) at the time of disclosure by the Company to the Consultant was in the public domain; (ii) after disclosure by the Company to the Consultant becomes part of the public domain through no fault of the Consultant; (iii) the Consultant can show as a result of written records was in his or her possession at the time of disclosure and was not acquired directly or indirectly from the Company; (iv) is independently developed by the Consultant without reference to or reliance upon Confidential Information as can be documented by written record; or (v) the Consultant is required to disclose pursuant to an order of a judicial or governmental authority, *provided* that the Consultant uses reasonable efforts to prevent such disclosure, and *provided further* that if the Consultant is unable to prevent such disclosure, the Consultant will provide the Company with written notice reasonably prior to such required disclosure so that the Company may seek to prevent or limit such disclosure."

Sometimes the agreement will allow the company to obtain an injunction in the event of a breach of a confidentiality obligation.

4. Access to Your Manuscripts

Consulting agreements sometimes call for the academic consultant to submit to the company all manuscripts before publication, including manuscripts arising strictly from research in the academic laboratory. To us there is much at risk to agreeing to this term. Obviously university policies might prohibit sharing manuscripts with third parties. Furthermore, doing so might complicate or impair patent applications. If the consultant is only providing advice to the company and engaging in no active research in its behalf, then the consultant should be able to convince the company that the company has no need to review manuscripts of company research—since there will be none—or of work from the academic laboratory. Alternatively, if the company limits the provision to company research only, there might be little harm in agreeing not to publish because, as noted above, there will be no company research anyway.

What is key, then, is not to commit to share manuscripts of work arising from the academic laboratory.

This situation is more complicated if the professor is not only consulting for the company but also receiving sponsored funding or is otherwise collaborating with the company concerning research in the academic laboratory. See Chapter 12. In that situation it is useful to view each relationship and each contract separately, especially when one considers that the overall situation can change over time and one relationship might end while another continues. So, although manuscript review might be appropriate in the sponsored funding agreement, it could properly be excluded from the consulting contract if the consulting work will be advisory only.

8 | IP Rights

IP RIGHTS ARE A KEY PROVISION OF every consulting agreement, and in some cases there might be nothing of greater potential importance to you in a consulting agreement than the IP provisions. Improperly surrendering your rights in IP, whether current discoveries or those you make in the future, could adversely affect a core function of a university laboratory, impair its academic freedom, and of course reduce the opportunity of the principal investigator and others on the principal investigator's team to receive royalties and strengthen the university's financial condition. In addition, issues concerning IP are becoming even more important following the recent decision by the U.S. Supreme Court in the *Stanford v. Roche* case, which, as discussed in Chapter 1, is likely to create greater challenges for universities in protecting research developed in their laboratories. IP rights provide another opportunity for the parties to set expectations appropriately.

The company will pay great attention to its IP rights in the consulting contract. You are responsible for ensuring that the company's language does not threaten your IP rights or those of your university, your collaborators, other consultee companies, and anyone else to whom you owe obligations.

The company will believe, reasonably, that it should have ample rights to the work you furnish to them. But the company's attorney or paralegal who is working on the contract might not appreciate that your consulting services are only a portion of your overall professional activities, that your primary responsibility is (presumably) to the university, or that you also might have obligations to other companies or third parties. So the IP provision as drafted by the company might be broader than what is reasonable given your circumstances.

Even if the contract is done properly, you should be alert to any other documents the company might ask you to sign relating to IP, including assignments. A key lesson of *Stanford v. Roche* is that an inventor's assignment to a company might trump your obligations to the university.

A. Ownership Rights

Sometimes the company's IP provision will assert rights to all of your work, wherever the work is done. Or the company might send you a patent or IP assignment that might be unjustifiably broad. Indeed, the nature of assignments and the complexities of patent rights encourage a broad approach, and no matter who benefits from an IP assignment—whether it is a company or your university—expansive language is to be expected and often justified. The risk, of course, is that your conveyance of IP rights to the company, regardless of whether you do so via a consulting agreement or an invention assignment, could deprive the university (and you) of royalties if not drafted appropriately. There is a special risk if inventions arise partly from work at your laboratory and partly from work at the company. (This comment assumes that your university would allow industrial consulting concerning topics being studied in your academic laboratory. We do not endorse the wisdom of such activities unless all parties—especially you and the university—understand the possible risks and accept them. See further discussion in Chapter 12.)

Therefore, an often preferable IP provision excludes rights to university inventions or discoveries and preserves the priority of the university's claims in the event of a dispute. An example follows:

> ### Example
> "The Consultant hereby assigns to the Company any right, title, and interest she may have in any invention, discovery, improvement, or other intellectual property that (i) the Consultant develops as a result of performing consulting services for the Company under this Agreement and (ii) is not developed in the course of Consultant's activities as a University faculty member ('Company Inventions')."

The company might not like this language, especially if the consulting work will focus on or overlap with your laboratory studies. Nonetheless, it is important to protect the university's (and your own) rights, and, as noted above, to be particularly careful considering the risks of acting in conflict with policies of your university administration.

The policies of some universities might require even tighter language to erect a higher bar for the company to surmount when claiming ownership of your IP and to strengthen the university's position in the event of a tug of war arising from overlapping claims to the same discovery. For this reason, an institution with this concern might welcome the company's IP rights being limited to inventions developed "solely" as a consultant and/or only those that are a "direct" result of performing consulting services.

Example

"The Consultant hereby assigns to the Company any right, title, and interest she may have in any invention, discovery, improvement, or other intellectual property which (i) the Consultant, alone or with others, develops solely as a direct result of performing consulting services for the Company under this Agreement and (ii) is not developed in the course of Consultant's activities as a University faculty member ('Company Inventions')."

Again, check your university's policies.

As noted, some companies might resist either of the examples presented above. They might argue that the language will unfairly deprive the company of credit for its share of work that is partly done for the company and partly for the institution, or that it puts the company in a position where, because it cannot monitor what you do in your laboratory, it might not own IP for reasons beyond its knowledge or control. The first situation—joint ownership—is unlikely in most cases. If your consulting work does not relate directly to your current academic research and the contract prevents you from engaging in active research for the company or using university personnel or resources, there should be no overlap. The second situation—the company's inability to monitor your academic work—is inherent when companies hire academics as expert advisors. Hopefully a company will agree to your (or your university's) point of view in both these instances. If, however, the consulting work is tied to your academic research, then the situation can be perilous, and certainly you should not undertake the assignment without vetting and approval of your university and a full understanding by you of the potential risks, including loss of your share of university royalties arising from inventions to which rights are conveyed to the company.

Note that consulting contracts from companies located outside the United States tend to have more aggressive IP claims than those from domestic companies. Not only are contracts overseas often written more loosely, but also non-U.S. companies might be unfamiliar with legal obligations that faculty in the United States owe to their institutions or to the NIH or other organizations. IP claims are another reason why you might wish to insist on having U.S. laws and the U.S. court system apply to the contract (discussed further in Section 14.B).

In the end, there is no easy solution when a company refuses to negotiate overreaching IP terms, especially if the company is unfamiliar with U.S. academic obligations. The company sometimes can be persuaded if the situation is put in perspective (e.g., the risk to the company in the circumstances is low; the university employs you and cannot oversee what you do for the company; the consulting job is expected to take, say, five days annually but you work for the institution throughout the rest of the year). However, a simple statement that no consulting can take place until the contract satisfies the university's policies is likely best.

Note also that the first example above does not address IP obligations owed to third parties other than the university, such as other companies you are advising. If you have such a concern, you can propose excluding inventions outside the Field that are not developed in the context of the consulting services. A company might be willing to accept the proposal in this light.

B. "No Infringement" Covenant

As also discussed in Chapter 8, a covenant (promise) in some biomedical consulting agreements is a statement from the consultant that all IP that she will generate for the company in the course of her consulting services will not infringe any third-party patents. We believe that it is inappropriate to require an academic scientist to agree to such a statement. It is in effect a guarantee and one that no one other than perhaps a registered patent agent can make. We do not know how anyone who believes that she has created a novel invention can guarantee that no one else has already made the same discovery and patented it. We believe that the company's patent experts are far better suited than the consultant to address this issue. Thus, we believe that this sort of provision should be deleted or at least be limited to the consultant's actual knowledge without any duty to know or assume the contents of other patent filings.

C. "Works for Hire" and "Moral Rights"

Sometimes the company's IP provision will say that your IP will be considered a "work for hire." It might also include your waiver of or assignment to the company of your "moral rights" in your IP.

These are both legal terms pertaining to copyrightable works, such as works of art or software code. The rule under the copyrights laws is that a copyrightable work is the work of its author. The exception is when the author has contracted with another party, such as an employer, to treat it as a "work for hire." A work for hire is owned by the employer, and the employer has all rights to exploit, modify, or claim credit for it.[1] So if the contract says that all IP you create in the course of your consultancy will be a "work for hire," then any copyrightable work will be the company's alone. This situation might give the company a presumption of ownership in court for work that might only partly be theirs if it is mixed with work you do at the university and therefore could make things harder for the university to protect the work done there. Our experience has been that some companies will strike the "work for hire" language if asked, especially if the consulting work is outside the software area. (Unlike other biotech IP, which usually is covered by patent laws, software is treated as copyrightable work.) You should also check university policy; there might be prohibitions that would help bolster a request to delete the language in a proposed contract.

We note that a law in California (Section 3341.5(c) of the California Labor Code) appears to treat a consultant as a company employee if the consulting contract states that the consultant's works of authorship (such as software or other copyrightable IP) created for the company are works for hire. Bloggers and others have criticized the law because it appears to require California companies to pay unemployment taxes and incur other obligations for code writers or other freelancers whom they might hire for a project from time to time. But the law might also raise issues for a consultant who is prohibited from accepting outside employment or from pre-committing future IP as works for hire. Therefore, if university policies are implicated, it might be prudent to discuss with California counsel whether work-for-hire language should be excluded from any consulting contract you will be performing in California or for a California company, or if California laws otherwise might be relevant.

The term "moral rights," from the French *droits moral*, does not address morality in a religious or ethical sense but rather encompasses certain rights of an author of a copyrightable material. An author's moral rights include the right to attribution and the right to prevent the alteration or distortion of the work.[2] Similar to the work for hire concept, your waiver of moral rights gives the company the right to alter your work and to cite it without naming you. It is our experience that some companies will drop the waiver of moral rights if asked, especially outside the software context.

D. Power of Attorney

Consulting agreements often include a power of attorney provision. Under this provision you appoint the company as your agent to sign future patent applications relating to IP you might generate as a consultant. The power of attorney can apply to all IP to which the company can claim rights under the agreement—thus taking you out of the process entirely—or only to IP that you fail to assign to the company because of disability, unavailability, or simply an unwillingness to do so.

As one might expect, an inventor's unwillingness to sign patent documents can signify his belief that the company is claiming work belonging to someone else. Thus, before you agree to the terms of a power of attorney provision, you (and your university) might consider whether the power of attorney could give the company the right to control future IP that you or the university might not feel is theirs.

Companies might agree to reduce the extent of the power of attorney or agree to give you and the university advance notice each time they intend to sign something on your behalf, so you and the university have a chance to object.

Example

"The Consultant hereby appoints the Company as his attorney to execute and deliver any such documents on his behalf in the event the Consultant should fail or refuse to do so within a reasonable period following the Company's request; *provided*, *however*, that before the Company exercises its authority as attorney-in-fact in any particular case, it shall first have given the University and Consultant 60 days written notice, by certified mail (the University's copy being mailed to the University's Office of Technology Transfer, [address]), that the Company intends to exercise its rights as attorney-in-fact under this sentence, and such notice shall be accompanied by a description of the relevant Invention."

E. List of Existing Inventions

Sometimes companies ask you to attach to the consulting agreement a list of all your existing inventions. We have never understood the purpose of this request. If the purpose is to prevent existing inventions from being later claimed as company property, the list adds nothing, because patent filings and invention disclosures will demonstrate what was preexisting technology in the event of a dispute. And if that is the purpose, what would it mean if you omitted an invention from the list inadvertently? More importantly, it is questionable whether furnishing this kind of list is consistent with confidentiality obligations owed to your university or your collaborators concerning technology not otherwise already in the public domain. Perhaps the request is made so that the consultant cannot later assert that a company invention was actually the consultant's prior invention, but even so there remains the question of what happens if the consultant is not scrupulous about identifying every past invention.

Fortunately, it is our experience that companies usually have no objection to dropping this disclosure requirement from the agreement.

9 | Noncompetition

A S ALREADY MENTIONED, CONSULTING contracts often impose limits on a consultant's competitive activities. You will need to consider whether the contractual terms are fair in light of your actual or anticipated obligations to the university or other companies.

One way to handle a noncompetition clause is to say that you do not agree to them in any context—in other words, that you are not promising that your consulting will be in any way exclusive to the company. Taking this position can succeed if the company really wants your services or if the company does not feel threatened by the possibility of your helping a competitor. The matter can be one of degree, that is, you and the company can control the breadth and duration of your obligation.

The best way to create a noncompete that you can live with is to fine-tune the scope of the "Field" to be very specific about the consulting topics to be performed. The narrower you can make the Field the less likely you will need to go back to the company to get a waiver to do other consulting. See Chapter 4.

Also, university policy might set an absolute bar by which nothing you do for a company may impinge on university research or the university's proprietary interest in it. In this context, preemptive language can be considered as follows:

Example

"The Company acknowledges and agrees, however, that nothing in this Agreement shall affect the Consultant's obligations to the University, research on behalf of the University, or research collaborations on behalf of the University in which the Consultant is a participant. Transfers of materials and/or intellectual property developed in whole or in part by the Consultant on behalf of the University shall not be affected by this Agreement."[1]

In looking at the above example, it is important to note not only that there is language to protect university research per se but also all interactions your laboratory has in scientific matters, even with other companies, such as scientific collaborations or material transfers.

As noted, be mindful of the definition of "Field" as well as "Consultant Inventions" to the extent they might reach your laboratory research either now or after the contract expires. You and the university do not want the contract to affect your unfettered pursuit of research as a faculty member.

Be mindful of a "tail period" that would prohibit you from consulting for anyone else in the "Field" for as long as it is in effect. Fortunately, as is the case with confidentiality clauses (see Chapter 7), the duration of a noncompete obligation often can be negotiated. Although a first draft of the consulting agreement might indicate that the obligation will last forever, usually a company will agree to limit the length to the life of the contract plus some defined tail period. We have seen contracts with no tail periods or with tail periods of between six months and five years. Three to five years is typical, but this range probably is not necessary and certainly not in your interest given the ever increasing pace of technological innovation. Because a noncompete clause could prevent you from consulting in the area in which you are most expert for a substantial period of time, you should seek the shortest possible term, ideally none, for any noncompete clause.

There is a belief among some in the business world that a noncompete provision in an employment contract (and likewise a consulting contract) is unenforceable and does not carry any weight. Putting aside doubts about such a broad contention, bear in mind that a primary mission of a contract negotiation—to establish the parties' expectations—is not met if one party is silent about what she thinks is an unreasonable noncompete clause. If you sign the contract, a company will believe you have promised to live up to the noncompete, and if you intend to ignore it then the stage is set for conflict. Therefore, putting legalities aside, to us there seems no reason to risk your reputation in the company's eyes. In sum, if the noncompete is too broad or too long, ask that it be fixed.

10 | Time Commitment

H ERE, THE FACTORS AT PLAY ARE how much time the company wants you to spend, how much you are willing to spend, what your other commitments are, and what university policy says.

University policy often limits the amount of time to be spent by faculty on outside activities such as consulting. We have already noted some of the more common time limits (see Section 2.E.1). You should be sure that the time commitment in the consulting contract does not exceed what the institution permits.

Even if the university has no policy, you might find it useful to set a time commitment. Doing so can help circumvent awkward discussions later if the company expects too much of you, and also makes it easier for you to determine how much time you will have available for additional opportunities and responsibilities. If a company comes back to ask for more of your time, you can reasonably request additional compensation. Also, if you are paid only via retainer (or stock) instead of by the hour or the day, then a time limit precludes the company from demanding so much time that the amount of your services end up being disproportionately high relative to your compensation.

Example

"Upon request by the Company, and at times mutually agreed upon by the Company and the Consultant, the Consultant shall devote up to ten days annually (each day to consist of up to eight hours) to providing consulting services to the Company pursuant to this Agreement."

11 | Term and Termination

A CONSULTING CONTRACT CAN BE FOR A defined period such as a year, or it can renew automatically unless one or both parties determine not to continue (called an "evergreen"). Consulting agreements for start-up companies often have three- to five-year terms and automatically renew for successive one-year periods unless the consultant or the company chooses to terminate the agreement.

The longer the term of a consulting contract the more important it is to consider giving either party the right to terminate the contract on a "no-fault" basis after a certain amount of time. Balanced against this flexibility is the desire of one or both parties to have the other locked in for as long as possible. For example, a consultant with unvested stock options might want to keep the contract going at least until the final vesting date.

For a multiyear contract, a company usually will agree to allow early "no-fault" termination by either party on some advance notice (say, 30 or 60 days).

Example

"This Agreement will have an initial term expiring on November 30, 2013, and thereafter will renew automatically for additional one-year terms unless either party notifies the other party in writing at least 30 days prior to the end of any one-year period that it elects not to continue the Agreement. In addition, at any time after November 30, 2013, either party may terminate this Agreement for any reason or no reason upon 30 days' prior written notice."

In connection with termination, consider the question of what provisions in the agreement will remain in effect thereafter (called "survival"). See discussion in Section 14.E.

A consulting contract also will usually permit the company to terminate it before the end of the term if you die, are disabled, or commit a "material breach" of the contract. Sometimes the material breach termination right will be made reciprocal, and you can ask for reciprocity if the first draft of the contract has it only going one way. If the company can terminate early, the contract will probably not permit you to exercise unvested options upon termination or get acceleration of subsequent fees unless you have very good bargaining leverage. However, on the flipside, it might be reasonable to have protections of this nature if it is the company that commits a material breach or other early termination event. See discussion of "Acceleration Events" in Section 6.C for details about how death or disability, a change of company control, or a material breach of the contract by the company can trigger the accelerated vesting of stock and stock options.

12 | Multiple Relationships with One Company

MULTIPLE RELATIONSHIPS BETWEEN A professor and a company are not uncommon. You might not only be a consultant but also a founder, a collaborator, or a board member. Before wearing a second or third hat, though, you should consider university policy. For example, some institutions do not permit faculty to consult for corporate sponsors of research in their laboratories or serve as corporate officers (see Section 3.D).

Next, be careful that a topic covered by one agreement with the company does not bleed into a topic covered by another agreement, and that if it does, the terms of the two agreements are not inconsistent. Here is an example. You agree to consult for a company that has licensed from the university some promising technology developed in your laboratory. The university provides the licensee with only very limited rights to add-on technology, because that is what its policy permits. If, however, you enter into a consulting agreement giving the company broad IP rights in the Field, then the company might claim from you the add-on technology that the university was barred from giving them under the license. Furthermore, the conveyance would be for free and deprive both you and the university of the opportunity to collect additional royalties. See also the discussion of manuscripts in Section 7.B.4.

With a start-up, your roles might be even more extensive: founder, director, shareholder, SAB member, general consultant, collaborator, inventor. You might also be a spokesperson or marketer. In all circumstances, it is important to fulfill your duties as a faculty member and ensure that you are complying with university policies. It is also important to confirm that the package of agreements you enter into with the start-up are consistent and reflect accurately the terms of the overall relationship.

13 | Start-Up Issues

As a founder of a start-up, you will probably have significant leverage to direct the terms of your deal, including such hot buttons as stock and stock options. The negotiation of an acceptable consulting agreement might be relatively easy, particularly if you are founding the company prior to involving any major investors. And if the university is an active participant, there might be no need for you to independently analyze university policies.

As a successful start-up evolves, a common early step is to acquire venture capital (VC) funding. VC firms will evaluate the start-up's stock structure and likely will ask for modifications to optimize their investment. Thus, you should not be surprised if the VC firm asks to renegotiate important terms of your consulting agreement—including founders' stock, stock options, acceleration rights, antidilution rights, and voting rights—before it is willing to invest. Your importance to the company's continued operations might be the most important factor in terms of what you keep or give up. Even in the best circumstances, antidilution rights are likely to go away—although in the long run, this loss may be offset by the value of bringing in a VC partner, which can be critical to a company's success.

The issues relating to start-ups can be complex and the precise wording is important. You might want to retain expert advice to guide you through this process to help protect your interests.

14 | Other Clauses

A. Indemnity

A company might ask you for an indemnity. An indemnity in consulting agreements typically cover any losses that the company might incur relating to your consulting services and ordinarily includes your promise to pay the company's legal costs and damages if it should be sued or found liable for claims made against the company arising from your services to the company.

A typical provision might read as follows:

> ### Example
>
> "The Consultant shall be responsible for, and shall indemnify and hold harmless the Company and its directors, officers, employees, and affiliates in respect of, any damage, claim, loss, or liability (whether criminal or civil and including any and all costs and expenses properly incurred in connection with any claim, including without limitation reasonable attorneys' fees and out-of-pocket expenses) of or suffered by the Company or any of its directors, officers, employees, or affiliates, or in settlement of any thereof, resulting or arising from the Consultant's performance of the Services under this Agreement or any breach of any of the Consultant's representations, warranties, or covenants hereunder, except to the extent that such damage, loss, or liability occurs from the gross negligence or intentional misconduct by the Company or the party seeking to be indemnified."

In essence, an indemnity of this nature makes you a guarantor of any future liabilities that the company might face arising from your services. But you might ask: what liabilities? After all, if you are simply giving advice, the company is free to listen or not. It seems a stretch to say that a company is at a significant risk of loss for the advice furnished to it by an academic consultant.

The company might argue that you should be responsible if you share with them information or ideas that infringe upon someone else's patent. But one response to this is to ask whether it is more reasonable to put the burden of this risk on you, an academic advisor, who might have no idea what is or is not patented, or on the company, which has patent counsel whose job it is to review patent filings.

By your being an indemnitor, the company can claim against your assets if it ever has reason to enforce the indemnity. Thus, to put it starkly, in exchange for a few days of consulting fees, one's net worth might be put at risk.

Needless to say, you might question whether an indemnity is fair. (In this context we are reminded of a colleague's remark: "My father always said if you have to indemnify a company, don't sign—period.") Fortunately, we have observed that companies generally are willing to forego the consultant indemnity because either the risk to the company is far-fetched or the potential penalty to the consultant is inappropriately severe. However, sometimes a company will not drop the clause entirely but instead insist on a more limited indemnity addressing perhaps one or two provisions of particular importance to the company (for example, the duty not to disclose confidential information). At the very least, one might argue, a company should agree to cap the amount owed under the indemnity to some reasonable amount, say, the total amount of fees paid, so that one's net worth is not exposed to indemnity claims under the contract.

It is important to note that getting rid of your indemnity does not mean the company cannot sue you later if it believes that your services caused the company to suffer a loss. A company can sue you with or without an indemnity. What the indemnity gives the company is the contractual right to pass along their legal fees and costs in defending actions brought by third parties relating to your services, and to have a claim against you under the contract for all damages or moneys paid in settlement of the claim. Moreover, even if you are certain you are not liable—say, the company itself was negligent—note that many indemnity provisions will require you to pay the company's legal fees and expenses as they are incurred and allow you to get them back only once a court has ruled in your favor, which might take years. To avoid exposure to a lawsuit from the company with unlimited risk of damages, some consultants ask that the contract cap the maximum amount they might ever owe the company for losses arising from the contract. As noted, that amount may be a dollar figure or the aggregate amount received from the company in the preceding calendar year, or some other simple formula. This provision often will also include a statement that the consultant will be liable only for actual out-of-pocket losses and not for punitive, compensatory, or exemplary damages. In addition, it can be prudent to look into buying insurance coverage of your potential liability to the company. Many liability or umbrella policies can be written to address this.

A Note about University Indemnities

We note that at least a few universities and research institutions will require a company to indemnify them for claims arising from consulting services furnished to the company by their faculty member. Their viewpoint is that they are not party to and do not benefit from the consulting relationship, and thus their risks of liability as your employer, however remote, should reside with the company rather than them. The obligation to indemnify the university should be considered separately from your indemnity of the company. In other words, the university's policies might require the contract to include a university indemnity, but that is independent of, and offers no basis for, whether you should be indemnifying the company in the contract. Fortunately companies usually understand and accept the distinction.

A Note about Company Indemnification of a Consultant

A company might be willing to indemnify you for actions or claims brought against you by third parties in connection with your services to the company. In our experience a company often will do so on request if the company is indemnifying the university or research institution pursuant to their policies (see immediately above). In other cases, however, unless you have agreed to indemnify the company, it might not be easy to convince them to give a one-way indemnity in your favor. However, an indemnity for you is worth pursuing if you are consulting concerning a matter for which the company might bear significant commercial risk, such as a clinical trial or for which your consulting could lead to products to be used by patients or members of the public.

If the company agrees to indemnify you, then the indemnity not only should guarantee the company's obligation to fully cover your losses and expenses but also should require the company to pay your expenses as they are incurred, so that you would not have to bear such expenses over possibly years of litigation with only a company IOU. Typically an indemnity provision will include language concerning the advancement of expenses; you should request such language if it is not there. An indemnity usually will not cover your gross negligence or willful misconduct or your material breach of the consulting contract. As noted, invalidation of an indemnification usually requires a final court finding of gross negligence, willful misconduct, or breach establishing an objective requirement to block indemnification.

Example (Pro-Consultant)

"The Consultant shall be indemnified and held harmless by the Company from and against any claim, demand, controversy, dispute, cost, loss, damage, expense (including, without limitation, attorneys' fees), judgment, and/or liability incurred by or imposed upon the Consultant in connection with any action, suit, or proceeding (including, without limitation, any proceeding before any administrative or legislative body or agency), to which the Consultant may be made a party or otherwise involved or with which the Consultant shall be threatened, by reason of this Agreement, including, without limitation, (i) any breach of this Agreement by the Company or any of its directors, officers, affiliates, agents, assignees, or delegees (collectively, the 'Company Group'), or (ii) any services, substances, products, devices, materials, or other tangible or intangible property that may be developed, used, sold, or transported by or on behalf of the Company Group, unless, and only to the extent that, the matter requiring indemnification is determined with finality by a proper court to be wholly the direct and immediate result of gross negligence or willful misconduct of the Consultant or of a material breach of this Agreement by the Consultant."

As with all terms, it is easier to get a one-way indemnity if you have strong bargaining power based on your importance to the company.

As discussed in Section 3.C, note that the company's obligation to indemnify you will be worthless if the company is bankrupt or cannot pay. Many personal liability or umbrella insurance policies can cover this kind of loss. You should check with your carrier (and confirm then and periodically thereafter that your carrier is itself financially sound). You can consider buying supplemental insurance to protect yourself against claims arising from your consulting work, as noted above. Factor the insurance cost into your consulting fees if the cost will be significant.

B. Governing Law; Legal Remedies

The first version of most consulting agreements will say that the laws of the company's home state or country will govern the agreement. It might also contain a "consent to jurisdiction" by which you agree to be sued in their locale in the event that any dispute arises between the parties.

As noted, you might not want your obligations governed by someone else's laws, and your university might not want unfamiliar laws to apply to your IP or confidentiality obligations to a company. This issue is not so important with a company from another state but it can be a concern when the company comes from another country, particularly one with laws written in a language that your lawyer or the university lawyer does not understand. Similarly, you might not want to consent to the jurisdiction of a distant place and incur the cost and burdens associated with doing so.

In this context, we are unsure what would happen if the laws governing the consulting contract would allow the company to claim ownership of IP that the university rightfully can claim under your IP agreement with the university or otherwise under their policies. It is best to avoid this situation entirely. If the company will not agree to have the contract governed by U.S. law, then a fallback position might be to state that in the event that there is a conflict between the agreement and your university IP agreement or its policies, the conflict will be determined by applying U.S. law.

Be aware that even if the company's first draft makes no reference to governing law or jurisdiction for disputes, the company might still be able to sue you in their home country, and you might not be able to dispute the court's power to hear the case. In that situation the court might apply its laws rather than yours.

Thus, if you have any concern about the company's location, you may want the contract to provide that all litigation will take place only in your state or city, and that your state's laws will govern the contract. This position is not unreasonable, given that the company is "coming to you" (at least figuratively) for advice, and you are employed by, and subject to the policies of, a research institution in the same location.

If the company objects, one compromise is for each party to agree that if it should sue the other party, the suit must be filed where the other party is located and the laws there will govern (subject to university IP rights, for example, always being determined by the laws of the university's state). This provision not only protects each party from being sued in an inconvenient or possibly inhospitable court but also has the salutary effect of encouraging negotiations before litigation.

A lawyer can be consulted about these issues, especially if there are complications or uncertainties.

A Note about Arbitration

A contract may provide for nonbinding mediation or arbitration of disputes. Nonbinding mediation involves the retention of a mediator to listen to the dispute and seek to resolve all or at least some of the issues before the parties go to court. Arbitration is a substitute for court. Often the process begins by the parties each selecting one arbitrator, and the two arbitrators picking a third, and then the parties present written and oral arguments to the arbitration panel on the matters at issue. The arbitrators then make a ruling, which usually has the full authority of a decision made by a court and usually is enforceable in court. The arbitration is governed by the rules of the arbitration association identified in the contract (such as the American Arbitration Association). Often the rules allow for submission of a wider variety of evidence and other information than what would be allowed in court. Depending on the rules at hand, each party pays half the costs of arbitration or the loser pays all.

Any unusual or important rules relating to arbitration can be addressed in advance in the contract. For example, it can be agreed that any patent disputes be arbitrated by lawyers admitted to the U.S. patent bar, or if you are confident that you would never litigate a dispute with the company without solid cause, the contract could say that the loser of the arbitration will pay the other party's legal fees and costs. (This clause would change the default rule in the United States, namely that each party bears its own fees and costs regardless of outcome.)

The "conventional wisdom" is that arbitration is cheaper than going to court. The conventional wisdom also holds that arbitrators are more likely than a court to produce a compromise decision, so that each party gets some of what they want rather than vindicating one side or the other. Nonetheless, some research institutions prefer arbitration, whereas others refuse to sign any contract with an arbitration provision.

We do not have sufficient experience either to confirm or deny the conventional wisdom about arbitration. We can offer only one observation, and it is that the likeliest hot button for arbitration under your consulting contracts will be disputes between the company and the university over IP ownership, and if conventional wisdom is at all correct, there is a risk that arbitrators will not fully enforce your institution's claims. With that being the case, if you are presented a consulting contract with an arbitration clause, we suggest you make sure that your technology office has no objection. If there is a problem, then ask the company to exclude IP disputes from arbitration.

C. Use of Consultant's Name

There might be university restrictions on a company's use of your name or likeness, or the university's name. (See Section 2.E.9.) Check the policy.

Regardless of the policy, though, there might be reasons why you want to control the use of your name by the company for promotional purposes. There are various ways to address this issue in the contract. Examples include the following:

> ### Example (Strict Prohibition)
> "The Company may not use the Consultant's name or likeness for any purpose except to the extent that disclosure of the Consultant's name is required by law."

> ### Example (Prior Consent)
> "The Company may not use the Consultant's name or likeness for any purpose except (i) with the Consultant's prior written consent or (ii) to the extent that disclosure of the Consultant's name is required by law."

Example (Defer to Institutional Policy)

"The Company may use the Consultant's name, and in doing so may cite the Consultant's relationship with the University, so long as any such usage complies with the University's policies published at www._____."

Example (Promotional Use)

"The Company may use the Consultant's name, and in doing so may cite the Consultant's relationship with the University, so long as any such usage (i) is limited to reporting factual events or occurrences only, and (ii) is made in a manner that could not reasonably constitute an endorsement of the Company or of any Company program, product, or service. However, the Company shall not use the Consultant's name or the University's name in any press release, or quote the Consultant in any company materials, or otherwise use the Consultant's name or the University's name in a manner not specifically permitted by the preceding sentence, unless in each case the Company obtains in advance the University's consent, and, in the case of the use of the Consultant's name, the Consultant's consent as well."[1]

D. Consulting for Affiliated Companies; Assignment

A company might want you to consult both for the company and its affiliates. If this concept is mutually agreeable, then you would want to be sure the contract addresses matters such as time commitment and compensation in a manner that is inclusive of all consulting duties. You might also want to consider having the affiliates sign the contract as additional parties so, if ever necessary, you are able to enforce obligations against them directly, for example, if the affiliate is sold and then unilaterally reduces your fee. There might also be a university policy requiring affiliates to sign the agreement. The university might want such signatures so that there is no doubt about each affiliate's being bound by limitations in the agreement relating to your university research or IP.

One way to address this issue in the contract is as follows:

Example

"The Consultant shall consult for the Company and, with the Consultant's prior written consent, any affiliate or subsidiary of the Company that has agreed in writing for the benefit of the Consultant to be bound by the terms of this Agreement as if it were the Company identified herein."

A more limited way to address it is as follows:

Example

"The Consultant shall consult for the Company and its affiliate XYZ Company to the extent that XYZ Company has agreed in writing for the benefit of the Consultant to be bound by the terms of this Agreement as if it were the Company identified herein."

Be aware that there are ways that the contract can require you to consult for affiliate companies without your knowing about it.

One is in the definition of "Company" at the beginning of the agreement. This definition can be written in a way so that the "Company" for purposes of the agreement includes other entities as well.

Example

"This Agreement is between Dr. J. F. Smith and Acme Corporation (*together with its affiliates and subsidiaries*, the 'Company')." [emphasis supplied]

By defining the "Company" this way, all consulting obligations flow to all entities covered by the definition.

Another way this issue can be addressed is in the provision dealing with "assignment." Assignment is the process by which a party may transfer all or some of its contractual rights or duties to others. Unless the contract speaks otherwise, a party is free to assign any of all of its obligations to someone else. That is why consulting contracts generally will say that the consultant cannot assign his or her obligations, which is reasonable considering the company is hiring you specifically and does not want anyone else assuming your duties. Although assignment by the consultant is barred, typically a company will have the right to assign the consulting agreement freely. In this way, the company might elect to give some or all of its rights to its affiliates.

Example

"The Company may assign this Agreement, in whole or in part."

If this seems objectionable to you, there are other approaches. One is to preclude assignment except to a successor that acquires the company, or if assignment is required by law (such as in a bankruptcy).

Example

"The Company may assign this Agreement to any legal successor of all or substantially all of the Company's assets by merger, sale, bankruptcy, insolvency, or other transfer."

To restrict the company from sharing your services with affiliates or third parties, you might ask not to permit assignment "in part." This way you are required to furnish services only to the company or its legal successor, without having to accept assignments from other entities as well.

Example

"The Company may assign this Agreement wholly but not in part."

Finally, if you have the bargaining power, you might want to prohibit any assignments by the company. That way, the company and its assignee will have to come to you for permission to assign the contract. This limitation gives you control over where your services will be provided. Such a prohibition can insulate you from having to consult for someone who fails to meet your requirements as to, say, business ethics, or whom you fear will not be able to continue to pay your consulting fee. It also gives you a new opportunity to negotiate for better terms.

Example

"This Agreement shall inure to the benefit of and be binding upon the respective heirs, executors, successors, representatives, and assigns of the parties, as the case may be; *provided, however,* the obligations hereunder of each party to the other are personal and may not be assigned without the express written consent of the other party."

Typically assignment provisions are not controversial unless the company intends to have more flexibility in how your services are allocated. If that is the case, then it is better to find that out beforehand and work out something that is acceptable to both of you. Again, setting expectations can eliminate potential problems down the road.

E. Survival

An agreement can terminate while there are still remaining obligations of the parties. You might be owed fees, for example. The agreement will usually specify provisions that the company wishes to continue after termination. Such a provision is called the "survival" provision, meaning that the identified contractual terms will "survive" or continue after termination. Survival might be for a set amount of time or be indefinite. Typically the company will have your nondisclosure, noncompetition, IP, and indemnity (if any) obligations survive for a tail period or indefinitely. As already noted, it is in your interest to keep tail periods as short as possible or to remove them.

You should review the contract to make sure that the company's list and the tail periods are appropriate and to see what terms you would want added. Typically a consultant would want the following provisions to survive:

- Accrued but unpaid fees
- University's rights with respect to the consultant's IP
- Use of consultant's name
- Company indemnity of consultant (if any)

Also, if your institution has a policy requiring the company to indemnify them, then the institution will likely want this indemnity also to be included with the other provisions surviving termination.

F. Independent Contractors

The agreement usually will state that you and the company are "independent contractors." This phrase is put in contracts to preclude you from claiming that you are the company's agent, or vice versa.

Since many universities state that faculty consult as individuals and not as institutional representatives, you might also want to say that in the independent contractor provision.

G. Representations and Warranties—Your Guarantees

Sometimes a company will ask the consultant to make certain assurances to the company under the agreement, such as the consultant's stating that he or she is permitted to enter into the consulting agreement under university policies. These assurances are usually called "representations and warranties."

> *Example*
>
> "Consultant represents and warrants that she is not under any preexisting obligation, and will not assume any obligation during the Term, in conflict or in any way inconsistent with the provisions of this Agreement. Without limiting the foregoing, Consultant represents and warrants that she has obtained all approvals as may be required by her university in order to furnish the consulting services contemplated by this Agreement, and nothing in this Agreement is inconsistent with her obligations to the university."

If a company is asking you for "reps" or warranties, you should be sure that they are true. This might sound like a silly concern, and for many things it is. As cited earlier, it should be fairly easy to confirm that your institution allows you to consult and to confirm that you have satisfied any disclosure or consent requirements that it imposes upon faculty who wish to consult.

But frequently there are reps and warranties that you will know you cannot furnish, or upon a little thought will appear invalid or unclear.

A good example is a pretty common one. The first draft of the consulting contract will state that you will not use or infringe upon any third-party patents in the course of providing consulting services.

> *Example*
>
> "Consultant represents and warrants that she has the right to disclose and/or or use all ideas, processes, techniques, and other information, if any, that she has gained from third parties, and that she discloses to the Company or uses in the course of performance of this Agreement, without infringing any patents of third parties and otherwise without liability to such third parties. Further, Consultant agrees that she shall not bundle with or incorporate into any deliveries provided to the Company herewith any third-party products, ideas, processes, or other techniques, without the express, written prior approval of the Company."

This sounds easy to promise, but is it? Actually, it is not. As noted, academic scientists can be expected to know about the leading developments in their field but might not know which ones are patented or have other IP restrictions, and so academics typically should not be expected to function, in essence, as a patent agent by verifying what legal rights (or lack of legal rights) are attached to a piece of information he might or might not discuss with the company. On the other hand, companies do employ patent lawyers and patent agents and are better suited to determine this information.

15 | Use of Consulting Entity

IN OTHER AREAS, WE ARE FAMILIAR WITH individual consultants setting up wholly owned companies and contracting their services through these wholly owned companies. So, for example, Jane Smith may establish Smith Consulting, LLC, and Smith Consulting, LLC would sign consulting contracts. The main reasons for doing so would be that the separate entity could shield the person from personal liability arising from the contracts (although this is not bulletproof) and that there might be advantages under tax or pension laws.

16 | Conclusion

F OR THE MOST PART, FACULTY CONSULTING agreements are pretty straightforward. Nonetheless, given the legal obligations involved, the need to comply with university policy, and the importance of guarding against possible negative outcomes, it is not sensible to enter into these agreements in ignorance. The cost in time or in legal fees to act prudently can be insignificant compared with what might be gained or lost. A careful review of a consulting agreement usually will produce better terms, set the relationship with the company on a solid basis, and can save you from a contract that could put you or your laboratory research, your relationship with your university, or even your net worth at risk.

Basic Consulting Agreement[1]

THIS IS A SAMPLE OF A SIMPLE CONSULTING agreement with no equity component. It is provided for illustrative purposes only. Do not rely on this sample for completeness or accuracy.

Consulting Agreement

This Agreement is made as of _____, 20____, between [insert full legal name of company] (the "Company") and [insert name of consultant] (the "Consultant"), a faculty member and an employee of [insert name of university or other institution] (the "University"). The Company is engaged in scientific research in [insert a detailed description of the area in which the consultant will consult] (the "Field"). The Consultant has extensive experience in the Field, and the Company seeks to benefit from the Consultant's expertise by retaining the Consultant as a consultant. The Consultant wishes to perform consulting services in the Field for the Company. Accordingly, the Company and the Consultant agree as follows:

1. *Services.*
   (a) The Consultant shall provide consulting services to the Company with respect to matters related to the Field (the "Services"). The Services for the Company shall consist only of the exchange of ideas and provision of advice; the Consultant shall not direct or conduct research for or on behalf of the Company.
   (b) Upon request by the Company, and at times mutually agreed upon by the Company and the Consultant, the Consultant shall devote no more than [insert number] days (each day to consist of up to ___ hours) annually to providing consulting services to the Company pursuant to this Agreement.

(c) The Company acknowledges that the Consultant is a University employee and is subject to University's policies. This Agreement is subject to the University's policies as they may exist from time to time, and anything in this Agreement to the contrary shall be void.

2. *Compensation.* The Company shall pay the Consultant $ ___ per day of consulting (to cover up to __ hours of Services, or any smaller increment thereof actually performed), plus an additional fee of $ __ per hour for each additional hour of consulting services beyond __ hours furnished during any consulting day. Such payment shall be made promptly following the Company's receipt of the Consultant's invoice in the Company's required form.

3. *Competition.* The Consultant represents to the Company that the Consultant does not have any agreement to provide consulting services to any other party, firm, or company on matters relating to the Field, except for agreement(s) between the Consultant and [insert name(s) of other company or companies]. Subject to Section 1(c) above, during the term of this Agreement, the Consultant shall not provide consulting services to any entity other than the Company and [insert name(s) of other company or companies] on matters relating to the Field. The Company acknowledges and agrees, however, that nothing in this Agreement shall affect the Consultant's obligations to the University, research on behalf of the University, or research collaborations on behalf of the University in which the Consultant is a participant. Transfers of materials and/or intellectual property developed in whole or in part by the Consultant on behalf of the University shall not be affected by this Agreement.

4. *Confidentiality.*
   (a) The Consultant may disclose to the Company any information that the Consultant would normally freely disclose to other members of the scientific community at large, whether by publication, by presentation at seminars, or in informal scientific discussions. However, notwithstanding anything in this Agreement to the contrary, the Consultant shall not disclose to the Company any information that is confidential to the University or any third party.
   (b) In providing consulting services to the Company pursuant to this Agreement, the Consultant may acquire information that

pertains to the Company's Confidential Information (as defined below). The Consultant agrees, for a period of _____ years following her first receipt of such Confidential Information, not to disclose or make use of, or allow others to use, any Confidential Information, except to the Company's employees and representatives, without the Company's prior written consent; *provided, however*, that in any event all such confidentiality obligations will expire no later than _____ years after the termination or expiration of this Agreement.

(c) As used in this Agreement, "Confidential Information" shall comprise all trade secrets and confidential or proprietary information (and any tangible representation thereof) owned, possessed or used by the Company or learned of by Consultant in connection with the consulting services hereunder, in each case to the extent that each such item is marked by the Company prominently as "confidential" upon first presentation to Consultant, or if first presented to Consultant other than in writing, is reduced to writing and marked by the Company prominently as "confidential" and presented to Consultant within ten days after first disclosure, including Company Intellectual Property as defined below. The term "Confidential Information" hereunder shall not include information that the Consultant can establish by competent written evidence that (i) at the time of disclosure by the Company to the Consultant was in the public domain; (ii) after disclosure by the Company to the Consultant becomes part of the public domain through no fault of the Consultant; (iii) the Consultant can show as a result of written records was in his or her possession at the time of disclosure and was not acquired directly or indirectly from the Company; (iv) is independently developed by the Consultant without reference to or reliance upon Confidential Information as can be documented by written record; or (v) the Consultant is required to disclose pursuant to an order of a judicial or governmental authority, *provided* that the Consultant uses reasonable efforts to prevent such disclosure, and *provided further* that if the Consultant is unable to prevent such disclosure, the Consultant will provide the Company with written notice reasonably prior to such required disclosure so that the Company may seek to prevent or limit such disclosure.

5. *Intellectual Property*. The Consultant hereby assigns to the Company any right, title, and interest she may have in any invention, discovery, improvement, or other intellectual property that (i) the Consultant develops as a result of performing consulting services for the Company under this Agreement and (ii) is not developed in the course of Consultant's activities as a University faculty member. Any intellectual property assignable to the Company pursuant to the preceding sentence is hereinafter referred to as "Company Intellectual Property." Upon the request of the Company, the Consultant shall execute such further assignments, documents, and other instruments as may be necessary to assign Company Intellectual Property to the Company and to assist the Company in applying for, obtaining and enforcing patents or other rights in the United States and in any foreign country with respect to any Company Intellectual Property. The Company will bear the cost of preparation of all patent or other applications and assignments, and the cost of obtaining and enforcing all patents and other rights to Company Intellectual Property.

6. *Term and Termination*.
   (a) Unless terminated earlier under Section 6(b), below, this Agreement shall be for a term of [insert term of years/months/days as appropriate or a specific end date].

   (b) Without limiting any rights which either party to this Agreement may have by reason of any default by the other party, each party reserves the right to terminate this Agreement at its convenience by written notice given to the other party. Such termination shall be effective upon the date not earlier than 30 days following the date of such notice as shall be specified in said notice.

   (c) Sections 2, 4(b), 4(c), 5, 7(c), 7(g), and this Section 6(c) hereof shall survive termination of this Agreement indefinitely or as otherwise provided therein.

7. *Miscellaneous*.
   (a) This Agreement shall inure to the benefit of and be binding upon the respective heirs, executors, successors, representatives, and assigns of the parties, as the case may be; *provided, however*, the obligations hereunder of each party to the other are personal and

may not be assigned without the express written consent of the other party.

(b) The relationship created by this Agreement shall be that of independent contractor, and the Consultant shall have no authority to bind or act as agent for the Company or its employees for any purpose.

(c) The Company may use the Consultant's name, and in doing so may cite the Consultant's relationship with University, so long as any such usage complies with the University's policies published at www._____.

(d) Notice given by one party to the other hereunder shall be in writing and deemed to have been properly given or paid if deposited with the United States Postal Service, registered or certified mail, addressed as follows:

[Name and Address of the Company.]

[Name and Address of the Consultant.]

(e) This Agreement supersedes all prior or contemporaneous agreements and discussions relating to the subject matters hereof and constitutes the entire agreement between the Company and the Consultant with respect to the subject matters of this Agreement. [Without limiting the foregoing, this Agreement replaces and supersedes in its entirety the Confidential Disclosure Agreement between the parties dated as of _____ (the "CDA"). The CDA is hereby terminated in all respects.]

(f) The Company and the Consultant agree that any amendment of this Agreement (including, without limitation, any extension of its term or any change in the consideration set forth above to be provided to the Consultant hereunder) or any other departure from the terms or conditions hereof must be signed by the Consultant and an authorized representative of the Company.

(g) If any provision of this Agreement is adjudicated to be invalid, unenforceable, contrary to, or prohibited under applicable laws or regulations of any jurisdiction, such provision shall be severed and the remaining provisions shall continue in full force and effect.

(h) The Consultant and the Company acknowledge that (i) the Consultant is entering into this Agreement in the Consultant's individual capacity and not as an employee or agent of University, (ii) University is not a party to this Agreement and has no liability or obligation hereunder, and (iii) University is an intended third-party beneficiary of this Agreement and certain provisions of this Agreement are for University's benefit and are enforceable by University in its own name.

(i) This Agreement shall be governed by the laws of the State of _____.

IN WITNESS WHEREOF, the parties have executed this Agreement effective as of the date first stated above.

[Full legal name of Company]

[Note: Must be signed by an authorized representative.]

By: _____

Name:

Title:

Date:

_____

[Name of Consultant]

Date:

Glossary

Acceleration event: A future event—such as a change of control, a material breach, or a death or disability—which upon occurrence automatically accelerates the vesting of all as yet unvested stock or options.

AMT: Alternative minimum tax

BOD: A company's Board of Directors

Capital gain: A taxable event arising from the sale of property, such as stock. Under federal tax law, a capital gain can be short term (if the property is owned for one year or less), which at present is taxed at the ordinary income rate, or long term (if the property is owned for more than one year), which presently is taxed at a lower rate.

Carried interest: A percentage interest of a venture fund or other investment fund. For example, a 5% carried interest typically means a right to 5% of the fund's profits.

CDA: Confidential disclosure agreement; also sometimes called an NDA, for nondisclosure agreement (not to be confused with a new drug application)

CEO: A company's chief executive officer

Change of control: A term found in consulting or stock option agreements describing situations in which a company might change hands, e.g., a merger or sale of significant assets. A change of control might trigger certain rights on the part of the consultant (e.g., the right to terminate) or option holder (e.g., the right to accelerate vesting of options).

Charter: A company's organizational document, usually filed with the jurisdiction where it is organized. Corporate charters are usually called certificates or articles of incorporation.

COI: Conflict of interest

Common stock: Ownership interests (equity) in a company. Usually includes the right to elect directors, vote on other corporate matters, and receive dividends. Holders of common stock receive corporate assets after all other creditors, including those holding any preferred shares, in the event of the company's insolvency or sale.

Compensation income: Income for services; the recipient generally is obliged to pay self-employment taxes.

Covenant: An obligation applying to future behavior or actions

CSO: Chief scientific officer

Dilution: A decline (dilution) in the percentage ownership represented by previously issued shares, generally caused by a company's issuance of new shares

D&O insurance: Directors & Officers insurance policy, obtained by a company from an insurance company to insure its directors and officers in case of certain events, often including litigation

Equity: The term "equity" has different meanings based on context. For purposes of this book, "equity" means an ownership interest in a company, usually in the form of common stock.

Evergreen: A contract that automatically renews unless one party notifies the other(s) of its election to terminate

FDA: The U.S. Food and Drug Administration

FMV: Fair market value

Including without limitation: A phrase used in contracts to signify that a list of items is not comprehensive but only illustrative

IND: Investigational New Drug application

Indemnitee: A party entitled to be indemnified by an indemnitor

Indemnitor: A party obliged to indemnify an indemnitee

Indemnity: The obligation to pay certain costs, losses, damages, etc. of another person or entity

IP: Intellectual property

IPO: Initial public offering of a company's shares

IRS: U.S. Internal Revenue Service

ISO: Incentive stock option

Liquidation event: A significant corporate event for which a stockholder might be entitled to receive cash or other consideration. Liquidation events are established under the governing corporate documents or relevant contracts. Liquidation events might include some or all of a sale, lease, license, or merger of the assets of the business or an IPO.

LLC: Limited liability company. An LLC is a different legal entity from a corporation and offers more flexibility in structuring ownership. Owners of an LLC are called "members" rather than "stockholders."

Lock-up period: A period of time following a public offering during which stockholders commit not to sell their shares

NIH: U.S. National Institutes of Health

NQSO: Nonqualified stock option

Optionee: The holder of a stock option

Ordinary income: Income that is taxed at ordinary rates—to be contrasted with capital gains income, which is taxed at lower capital gain rates. Also called ordinary compensation income, which is income for which a payee who is not employed by the payer must pay self-employment taxes

Phantom income: Taxable income incurred without receipt of cash. The receipt of phantom income can require payment of tax before payment from the underlying asset and therefore can require having cash to pay the tax before the underlying asset has generated any income.

Preferred stock: An ownership interest (equity) in a company. Usually includes right to a preferred dividend (i.e., one that is payable before payment of dividends on common stock) and special voting rights on corporate issues. Stands ahead of common stock in the event of the company's insolvency or sale. Depending on the terms, preferred stock can be "convertible" into common shares either at a certain time or upon the election of the stockholder. A company can issue more than one class of preferred stock with different rights and privileges.

R&D: Research and development

Registration rights: A shareholder's contractual right to include his or her shares in a public or other offering of the company's shares

Restricted shares: Shares of a company's stock for which there are resale or transfer restrictions imposed under the terms of the agreement by which the holder of the shares acquired them. Restricted shares can be vested or unvested.

Restrictive legend: Language appearing on a stock certificate setting forth legal restrictions relating to the underlying shares. Such restrictions might identify vesting requirements pertaining to the shares or prohibit the transfer or resale of the shares.

SAB: A company's Scientific Advisory Board

SAR: Stock appreciation right

SEC: U.S. Securities and Exchange Commission

Section 83(b) election: A tax strategy to accelerate recognition of a taxable event in connection with the receipt of unvested shares. Generally, a properly filed Section 83(b) election accelerates the recognition of a taxable event from the date of vesting to the date of receipt.

Section 409A: A federal tax provision imposing certain penalties upon stock options or SARs issued below FMV

Stock options: The right to acquire shares of a company according to specific terms, including price (the "strike price")

Strike price: The price established in a stock option agreement at which an option holder can acquire shares underlying the option

Survival: The length of time that an agreed-upon contractual term will continue to apply after the contract has terminated

Tail period: A defined period of time after the termination of a contract during which a specific contractual right or obligation (such as the obligation not to disclose confidential information) will continue to be in force

Unvested shares: Shares for which the holder has not acquired full ownership rights

VC: Venture capital, venture capital funding, or a venture capitalist

Vested shares: Shares for which the holder has full ownership rights

Vesting: The process by which the holder of an interest in a security (such as a stock option or restricted shares) acquires full ownership rights in the security

Notes

Chapter 1

1. Board of Trustees of the Leland Stanford Junior University v. Roche Molecular Systems, Inc., no. 09–1159, 563 U.S. __ (2011), http://www.supremecourt.gov/opinions/10pdf/09-1159.pdf (hereafter *"Stanford v. Roche"*). For more discussion of *Stanford v. Roche*, see, e.g., C. Brinckerhoff, "Investor Rights not Affected by Bayh-Dole," *Genetic Engineering and Biotechnology News*, Aug. 1, 2011.

2. See T. Kaplan, "French Doctor Arrested on Insider Trading Charges," *The New York Times*, Nov. 2, 2010, http://dealbook.nytimes.com/2010/11/02/french-doctor-arrested-on-insider-trading-charges/. See also "Researcher Accused of Insider Trading Has Worked for Many Drugmakers," *FDA News*, Nov. 9, 2010, http://www.fdanews.com/newsletter/article?articleId=131743&issueId=14200; "Factbox: Who's Who in the Latest Insider Trading Arrests," *Reuters*, Feb. 8, 2011, http://www.reuters.com/article/2011/02/09/us-hedgefunds-idUSTRE7175 E920110209.

3. A scientist at the U.S. Food and Drug Administration was accused in March 2011 of trading on the basis of confidential results of drug-approval information learned in the course of his job. See A. Mundy and B. Kendall, "'Insider' is Charged at FDA," *The Wall Street Journal*, March 30, 2011, http://online.wsj.com/article/SB10001424052748704447190457623100091822187 0.html?mod=djemalert NEWS. Another case, from September 2011, involves a clinical physician-scientist at Columbia University who allegedly shared clinical results with a hedge fund. D. Wilson, "Columbia Professor is Linked to Insider Trading Case," *The New York Times*, Sept. 17, 2011, http://www.nytimes.com/2011/09/17/health/17insider.html?_r=1&scp=3&sq=insider%20trading&st=cse. See also Section 3.F of the main text. In addition, in early 2014 two university researchers testified that they had furnished confidential clinical trial data to a hedge fund executive who stood trial for insider trading violations. News reports indicated that the two testified in exchange for immunity from prosecution. See A. Stevenson, "Ex-SAC Trader's Drug Data a Surprise, Doctor Testifies," *The New York Times*, Jan. 15, 2014, http://dealbook.nytimes.com/2014/01/15/doctor-at-insider-trading-trial-testifies-it-was-like-he-was-in-the-room/.

4. J. Gravelle, "Illegal Insider Trading and the Healthcare Industry," *Federal Securities Law Blog*, http://www.fedseclaw.com/2009/05/articles/business-news-1/illegal-insider-trading-and-the-healthcare-industry/.

5. Department of Health and Human Services, "Supplemental Standards of Ethical Conduct and Financial Disclosure Requirements for Employees of the Department of Health and Human Services," August 31, 2005, 70 *Fed. Reg. 51559,* http://ethics.od.nih.gov/lawreg/5-CFR-5501-Unofficial-Compilation.pdf. For additional information, go to http://www.nih.gov/about/ethics_COI.htm. For background on the 2005 developments, see T. Agres, "NIH Eases Ethics Rules," *The Scientist*, Aug. 26, 2005, http://classic.the-scientist.com/news/20050826/01/.

6. "NIH Researcher Sentenced in Ethics Case," *The Washington Post*, Dec. 22, 2006, http://www.washingtonpost.com/wp-dyn/content/article/2006/12/22/AR2006122200676.html. See also "Conflict-of-Interest Inquiry May be Reopening at NIH," *The Washington Post*, March 31, 2007, p. A18, http://www.washington post.com/wp-dyn/content/article/2007/03/30/AR2007033000310.html.

7. See National Institutes of Health, "Financial Conflicts of Interest," http://grants.nih.gov/grants/policy/coi/.

8. "Regulators Probe Sale of Secrets by Doctors," *The Seattle Times*, Aug. 10, 2005, http://seattletimes.nwsource.com/html/localnews/2002430779_secinvest10m.html. See also "Senator Wants SEC to Probe Leaks of Drug-Research Secrets," *The Seattle Times*, Aug. 8, 2005, http://seattletimes.nwsource.com/html/health/2002426761_hedge08m.html; "Inslee Seeks Probe of Study Leaks," *The Seattle Times*, Aug. 13, 2005, http://community.seattletimes.nwsource.com/archive/?date=20050813&slug=hedge13; and "Med Schools Urged to Plug Drug-Research Leaks," *The Seattle Times*, Aug. 19, 2005, http://community.seattletimes.nwsource.com/archive/?date=20050819&slug=drugsecrets19.

9. *The Seattle Times* published a collection of articles covering this story under the lead story "Drug Researchers Leak Secrets to Wall Street," *The Seattle Times*, Aug. 7, 2005, http://seattletimes.nwsource.com/news/business/drugsecrets/. The other articles published in that edition of the newspaper are "Drug Researchers Leak Secrets to Wall St.," *The Seattle Times*, Aug. 7, 2005, http://seattletimes.nwsource.com/html/businesstechnology/drugsecrets1.html; "Some Doctors See Ethical Pitfall in Actions that Others Defend," *The Seattle Times*, Aug. 7, 2005, http://seattletimes.nwsource.com/html/businesstechnology/drugsecrets2.html; "Selling Secrets Can Distort Data, Kill Promising Drugs," *The Seattle Times*, Aug. 7, 2005, http://seattletimes.nwsource.com/html/businesstechnology/drugsecrets3.html; "Sellers, Buyers of Secrets Risk Being Prosecuted," *The Seattle Times*, Aug. 7, 2005, http://seattletimes.nwsource.com/html/businesstechnology/drugsecrets4.html; "Investors Quiz Researcher in Recorded Conference Call," *The Seattle Times*, Aug. 7, 2005, http://seattletimes.nwsource.com/html/businesstechnology/drugsecrets5.html; "How Wall Street Gets the Inside Scoop on Drug Testing," *The Seattle Times*, Aug. 7, 2005, http://seattletimes.nwsource.com/news/business/drugsecrets/howitworks.html; "Moving the Market," *The Seattle*

Times, Aug. 7, 2005, http://seattletimes.nwsource.com/news/business/drugsecrets/movingmarket.html; and "Good Questions Lead to Surprising Story," *The Seattle Times*, Aug. 7, 2005, http://seattletimes.nwsource.com/html/localnews/drugsecretsabout.html.

10. Thanks to Heidi E. Henning, General Counsel, Howard Hughes Medical Institute, for her Latin expertise.

Chapter 2

1. For more discussion of impairment to contract, see E. Farnsworth, *Contracts* (Little Brown, 1982), § 4.2 at pp. 213–214.

2. *Stanford v. Roche*, see Note 1 in Chapter 1.

Chapter 3

1. Stanford University, *Research Policy Handbook*, Section 4.3.2, "Definition of 'Consulting,'" http://doresearch.stanford.edu/policies/research-policy-handbook/conflicts-commitment-and-interest/consulting-and-other-outside#anchor-640. See also *"Stanford University Requirements for Faculty Consulting Activities and Agreements,"* Sept. 7, 2010, http://doresearch.stanford.edu/sites/default/files/documents/consulting_requirements_3.pdf.

2. For example, Del. Gen. Corp. Law § 145; N.Y. Bus. Corp. Law § 722.

3. For discussion of the "corporate opportunities" doctrine, see, for example, E. Welch and A. Turezyn, *Folk on the Delaware General Corporation Law: Fundamentals* (Aspen, 2005), §§ 141.2.8 and 141.2.9, at pp. 197–204.

4. Sarbanes-Oxley Act of 2002, P.L. 107–204. The full text of the Act is available at http://www.sec.gov/about/laws/soa2002.pdf. Many websites provide summaries of the law.

5. E.g., Securities Exchange Act of 1934, as amended, Section 16(b), 15 U.S.C. § 78p(b). For background information on Section 16 reporting and liability, the best reference is P. Romeo and A. Dye, *Section 16 Treatise and Reporting Guide* (Executive Press, 3d ed., 2008), also available at http://www.section16treatise.net/home/. Professor Peter Romeo and Alan Dye also maintain a website, http://www.section16.net/home/. Other reference materials include "Short-Swing Profits," at 4 CCH Fed. Sec. L. Rptr. ¶ 26,101, and J. Bromberg & L. Lowenfels, *Bromberg and Lowenfels on Securities Fraud and Commodities Fraud* (Clark Boardman, 1994). Rule 10b-5 under the Securities Exchange Act of 1934, as amended, is a general antifraud provision that also can be applied in connection with insider trading cases. State laws also often can apply.

6. Securities Exchange Act of 1934, as amended, Section 16(a), 15 U.S.C. § 78p(a). For more information, see the Romeo and Dye treatise and website cited in Note 5 above, as well as "Stock Ownership Reports by Insiders," 4 CCH Fed. Sec. L. Rptr. ¶ 26,071.

7. Securities Exchange Act of 1934, as amended, Section 16(b), 15 U.S.C. § 78p(b). For more information, see citations under Note 5 above.

8. See citations under Note 6 above.

9. See B. Mullins and S. Pulliam, "Hedge Funds Pay Top Dollar for Washington Intelligence," *The Wall Street Journal*, Oct. 4, 2011, p. A1, http://online. wsj.com/article/SB10001424053111904070604576514791591319306.html? KEYWORDS=expert+networks.

10. 950 CMR §§ 12.200 *et. seq.* See Adopting Release, Aug. 8, 2011, http://www.sec. state.ma.us/sct/sctnewregs/description_of_changes_to_proposed_regs.pdf. See also D. Wilson, "Columbia Professor is Linked to Insider Trading Case," cited in Chapter 1, Note 3.

Chapter 4

1. This language mostly is based on the Howard Hughes Medical Institute ("HHMI") model agreement, "HHMI Model Consulting Agreement Form SC 512," (hereafter called the "HHMI Model Consulting Agreement"), as published several years ago. As of press time, a revised version is available by activating the "Download HHMI's Model Consulting Agreement" under the Consulting heading at http:// www.hhmi.org/about/policies. This language and other examples used in the text come from a model consulting agreement published or previously published by HHMI. See the Acknowledgments.

Chapter 6

1. For example, the National Venture Capital Association has a useful reference page (http://www.nvca.org/index.php?option=com_content&view=article&id=108 &Itemid=136) that includes model documentation for preferred stock awards and other VC-related agreements.

2. For basics on ISOs, see, for example, http://www.investorwords.com/2396/ incentive_stock_option.html, http://www.investopedia.com/terms/i/iso.asp, and http://www.fairmark.com/execcomp/iso.htm.

3. For basics on NQSOs, see, for example, http://www.investorwords.com/3340/ non_statutory_stock_option.html, http://www.investopedia.com/terms/n/nso.asp, and http://www.fairmark.com/execcomp/nqo.htm.

4. For more information, see J. Gworek, "Double Trigger Acceleration: Neat in Theory, Messy in Practice" (March 2008), at http://www.mbbp.com/resources/ business/double_trigger.html; and M. Davis, "Acceleration Triggers" http://mpd. me/acceleration-triggers/ (subscription required).

5. For a basic discussion of piggyback registration rights, see http://www.investopedia. com/terms/p/piggybackrights.asp.

6. For more information about securities rules relating to the resale of unregistered shares of a public company, including the SEC's Rule 144, see, for example, U.S.

Securities and Exchange Commission, "Rule 144: Selling Restricted and Control Securities," http://www.sec.gov/investor/pubs/rule144.htm.

7. A query of any internet search engine should produce materials discussing the tax implications of NQSOs and restricted shares. One example is SmartMoney. com, "Taxes on Nonqualified Stock Options," http://prelive.smartmoney.com/personal-finance/taxes/taxes-on-nonqualified-stock-options-9304.

8. More discussion of Section 409A is found at the IRS website, http://www.irs.gov/Retirement-Plans/409A-Nonqualified-Deferred-Compensation-Plans.

Chapter 7

1. This language is based on paragraph 4(b) of the HHMI Model Consulting Agreement. See Chapter 4, Note 1.

Chapter 8

1. For more information about works for hire, see, for example, U.S. Copyright Office, Information Circular "Works Made for Hire," revised September 2012, http://www.copyright.gov/circs/circ09.pdf.

2. For more information on moral rights, see, for example, B. Rosenblatt, "Moral Rights Basics," March 1998, http://cyber.law.harvard.edu/property/library/moral primer.html.

Chapter 9

1. See Chapter 4, Note 1.

Chapter 14

1. See Chapter 4, Note 1.

Attachment A

1. See Chapter 4, Note 1.

Index